Schriftenreihe aus dem Institut für Strömu

Herausgeber
J. Fröhlich, S. Odenbach, K. Vogeler

Institut für Strömungsmechanik
Technische Universität Dresden
D-01062 Dresden

Band 12

Bernhard Vowinckel

Highly-resolved numerical simulations of bed-load transport in a turbulent open-channel flow

TUD*press*

2015

Die vorliegende Arbeit wurde am 19. September 2014 an der Fakultät Maschinenwesen der Technischen Universität Dresden als Dissertation eingereicht und am 20. Februar 2015 erfolgreich verteidigt.

This work was submitted as a PhD thesis to the Faculty of Mechanical Science and Engineering of TU Dresden on 19 September 2014 and successfully defended on 20 February 2015.

Gutachter | Reviewers
Prof. Dr.-Ing. habil. Jochen Fröhlich, TU Dresden
Prof. Alfredo Soldati, University of Udine

Bibliografische Information der Deutschen Nationalbibliothek
Die Deutsche Nationalbibliothek verzeichnet diese Publikation in der Deutschen Nationalbibliografie; detaillierte bibliografische Daten sind im Internet über http://dnb.d-nb.de abrufbar.

Bibliographic information published by the Deutsche Nationalbibliothek
The Deutsche Nationalbibliothek lists this publication in the Deutsche Nationalbibliografie; detailed bibliographic data are available in the Internet at http://dnb.d-nb.de.

ISBN 978-3-95908-002-6

© 2015 TUDpress
Verlag der Wissenschaften GmbH
Bergstr. 70 | D-01069 Dresden
Tel.: 0351/47 96 97 20 | Fax: 0351/47 96 08 19
http://www.tudpress.de

Technische Universität Dresden
Fakultät Maschinenwesen

Highly-resolved numerical simulations of bed-load transport in a turbulent open-channel flow

Dissertation

zur Erlangung des akademischen Grades
Doktor-Ingenieur (Dr.-Ing.)

von

Dipl.-Hydrol. Bernhard Vowinckel

geboren am

04.11.1982 in Essen

Tag der Einreichung: 19.09.2014

I choose to listen to the river for a while, thinking river thoughts,
before joining the night and the stars.

Edward P. Abbey, American author and essayist

Acknowledgements

First of all, I would like to thank Prof. Jochen Fröhlich for supervising my thesis. During the years of my PhD he taught me good scientific practice, which is ruminating the study a couple of times until it becomes a reader's digest. This can be exhausting at times, but I stood to gain from this in the long run. I thank Dr. Tobias Kempe for developing the code PRIME and for being an important contact person for big and small questions. I would like to thank Prof. Vladimir Nikora for introducing me into the Double-Averaging Methodology and all the valuable talks we had about turbulence in open-channel flows. A special thanks goes out to my colleagues at the coffee table of the ISM, as they provided a comfortable environment to work in and one or another word of wisdom. I would like to acknowledge the financial support of the DFG via the project DFG FR 1593/5 as well as the ZIH, Dresden, and the JSC, Jülich, for providing large amounts of computational time. Prof. Soldati is acknowledged for accepting the part as the second examiner of my thesis.

Ein besonderer Dank gilt meiner Familie. Zu aller erst danke ich meiner Mutter für moralische, finanzielle und manchmal auch fachliche Unterstützung. Meiner leiblichen als auch angeheirateten Familie gebührt ein besonderer Dank für die Unterstützung auf den letzten Metern, als diverse Krankheiten drohten, der Sache einen Strich durch die Rechnung zu machen. Meinen Kindern danke ich dafür, dass Sie mich auf Trapp halten und jeden Tag zeigen, worum es im Leben eigentlich geht. Meiner Frau, Juliane Vowinckel, möchte ich für alles danken, was Sie in den letzten fünf Jahren für mich getan hat. Sie war immer da, wenn ich Sie gebraucht habe und hat mich mit Rat und Tat beim Gelingen dieser Arbeit unterstützt. Herzlichen Dank!

Abstract

This thesis presents the analysis of phase-resolving Direct Numerical Simulations of a horizontal turbulent open-channel flow at small relative submergence, which is laden with a multitude of spherical particles. These particles have a mobility close to their threshold of incipient motion such that they are transported in bed-load mode by constantly colliding with the sediment bed. The coupling of the fluid phase with the disperse phase is realized by an Immersed Boundary Method.

The thesis provides a detailed study addressing the impact of the choice of collision model on the scenario of bed-load transport illustrating the necessity of a sophisticated representation of particle contact to reproduce patterns known from experimental evidence. Statistical tools are presented to identify and describe the key-mechanisms governing the fluid-particle interaction. In this thesis, the Double-Averaging Methodology is applied for the first time to the situation of mobile rough beds. Taking advantage of the highly-resolved datasets produced, this methodology provides a framework to convolute the data in such a way that the most prominent flow features are well described by a handy set of double-averaged (in time and space) quantities.

The thesis further provides a systematic study elucidating in detail the impact of the key-parameters mobility and sediment supply on the pattern formation of large-scale particle clusters. This is done using a very large computational domain to allow bed-forms to evolve with minimal spatial constraints. Similar to experimental observations, it is found that a low transport rate is linked to streamwise oriented ridges, while a large sediment supply results in large-scale clusters that propagate in streamwise direction. A detailed description of fluid quantities links the developed particle patterns to the enhancement of turbulence and ultimately to a modified hydraulic resistance. The large domain allows for a large number of independent erosion events, such that conditional averaging provides a very clear description of the processes involved for incipient particle motion. Furthermore, the detection of moving particle clusters as well as the investigation of their surrounding flow field is performed by an analysis using a moving frame coordinate system.

The presented numerical method as well as the statistical tools are shown to be very suitable measures to study the complex situation of bed-load transport in open-channel flow and to give detailed insight into the key-mechanisms of particle-laden flows.

Zusammenfassung

Die vorliegende Dissertation präsentiert die Analyse von phasenaufgelösten Direkten Numerischen Simulationen eines offenen Gerinnes mit geringer Wassertiefe, das mit einer Vielzahl von sphärischen Partikeln beladen ist. Die Mobilität der Partikel ist dabei nah am Grenzwert für den Bewegungsbeginn gewählt, so dass der zu erwartende Prozess als Geschiebetransport klassifiziert werden kann, d.h. die Partikel befinden sich ständig am Gerinnegrund. Die Kopplung der flüssigen und der festen Phase erfolgt durch eine Immersed Boundary Methode. Die Dissertation stellt eine ausführliche Studie zum Einfluss der Wahl des Kollisionsmodels auf das gewählte Szenario des Geschiebetransports vor. Es wird gezeigt, dass ein hochentwickeltes Model nötig ist, um den Kontakt von Partikeln mit hinreichender Genauigkeit zu beschreiben, damit Ergebnisse aus experimentellen Beobachtungen reproduziert werden. Statistische Werkzeuge werden vorgestellt um die Schlüsselmechanismen der Interaktion zwischen Fluid und Partikel zu identifizieren und zu quantifizieren. In dieser Dissertation wird die Double-Averaging Methodik das erste Mal auf die Situation einer turbulent Strömung über eine mobile, raue Sohle angewandt. Auf der Basis der hochaufgelösten Daten ist es möglich, mit dieser Methodik die großen Datenmengen auf wenige Größen, die in Raum und Zeit gemittelt wurden, zu verdichten, um die wichtigsten Eigenschaften der Strömung zu beschreiben.

Darüber hinaus stellt die Dissertation eine systematische Studie der Variation der Schlüsselparameter Mobilität und Sedimentzufuhr und die damit einhergehende Bildung von Partikelanhäufung vor. Dieses Vorhaben wird mittels eines sehr großen Rechengebiets erreicht, damit die Formation der Partikelstrukturen ohne räumliche Einschränkung entstehen kann. Ähnlich wie bei Experimenten mit höherer Reynoldszahl findet man für die gegebene Konfiguration bei geringer Intensität des Geschiebetransports kammartige Strukturen, die sich in Strömungsrichtung erstrecken. Bei höheren Transportraten bilden sich großskalige Partikelstrukturen, die sich in Strömungsrichtung bewegen. Eine detaillierte Beschreibung von Strömungsgrößen stellt diese Beobachtungen in den Zusammenhang mit der Verstärkung der turbulenten Fluktuationen und dem einhergehenden modifizierten hydraulischen Widerstand.

Durch das große Rechengebiet konnte eine hohe Anzahl an Erosionsereignissen realisiert werden, so dass eine konditionelle Mittelung dieser Ereignisse ein sehr klares Bild der involvierten Prozesse bei der Partikelmobilisierung liefert. Darüber hinaus wird ein Algorithmus vorgestellt, der es erlaubt, Partikelstrukturen in Raum und Zeit zu verfolgen, so dass das Geschwindigkeitsfeld des umgebenden Fluids durch eine Koordinatentransformation analysiert werden kann.

Zusammenfassend zeigt sich, dass die vorgestellte numerische Methode sowie die statistischen Werkzeuge sich als sehr nützlich erweisen, die komplizierten Prozesse des Geschiebetransport in einem offenen Gerinne zu beschreiben.

Contents

Nomenclature

Roman Symbols

a, b	parameters of linear regression
D	Particle diameter
D^+	Particle Reynolds number based on friction velocity
d_n	damping coefficient
e	restitution coefficient
$\mathbf{e}_i$	unit vector in the i-th direction
E	numerical efficiency (weak scaling)
$\mathbf{f}$	volume force
f_x	streamwise component of volume force
f_y	wall-normal component of volume force
Fr	Froude number
$\mathbf{F}_p$	forces resulting from particle-particle interaction
G	Cartesian distribution function
g	gravitational acceleration
$\mathbf{g}$	gravitational vector
g_n, g_t	relative normal and tangential velocity
h	grid cell size
H	channel height / water depth
H_0	statistical hypothesis
H_{bal}	ballistic height
$\mathbf{I}$	identity matrix
I_{reg}	interval of linear regression
I_p	moment of inertia
k	dimensioned parameter
K_i	integral constant
K_p	proportional constant
k_n	material stiffness
L_{slip}	slip length
L_{trans}	mean free path
L_x, L_y, L_z	domain length in x-, y-, and z-direction
$L_{0,x}, L_{0,y}, L_{0,z}$	length of averaging domain in x-, y-, and z-direction
l_τ	viscous lenght scale
$\mathbf{M}_p$	moments resulting from particle-particle interaction
m	mass

$\mathbf{n}$	surface normal vector with n_1, n_2, and n_3
N_l	number of Lagrangian marker points
N_x, N_y, N_z	number of grid cells in x-, y-, and z-direction
N_p	number of particles
N_{pairs}	number of particle pairs
N_{ref}	number of processores of the reference run
N_{tot}	total number of grid cells
p	pressure
P	coordinate of a given point
P_i	master or slave processor
P_x, P_y, P_z	number of processors in x-, y-, and z-direction
q	confidence interval
$\mathbf{r}$	position vecor of marker point
R_p	particle radius
R_{uu}	two-point correlation function
r_x, r_z	correlation space coordinates in x- and z-direction
Re	Reynolds number
Re_b	Reynolds number based on bulk velocity and channel height
Re_p	Reynolds number based on slip velocity and particle diameter
Re_τ	Reynolds number based on friction velocity and channel height
S	range of repulsive force
S	computational speedup (strong scaling)
s_b, s_r, s_{t*}	t-test variables
S_{int}	area of the fluid-particle interface
Sh	Shields parameter
St_c	Stokes number
t	time
$\mathbf{t}$	unit vector collinear with tangential force vector
T_0	averaging time
t_c	time of a collision
T_b	bulk time unit
T_c	collision time interval
T_f	time a given location is occupied by fluid
T_n	averaging time interval
T_{sample}	total sampling time
t_r	test statistics
u_τ	friction velocity
U_b	desired bulk velocity
$U_{b,a}$	actual bulk velocity
$\mathbf{u}$	velocity vector with Cartesian components u_1, u_2, u_3
$u_1 = u$	velocity component in x-direction
$u_2 = v$	velocity component in y-direction
$u_3 = w$	velocity component in z-direction
V_0	spatial averaging domain
$V_{0,1}$	averaging volume covering two full lengths of computational domain
$V_{0,2}$	averaging volume covering one full length of computational domain
$V_{0,3}$	averaging volume covering parts of computational domain
V_l	volume associated to a Lagrangian marker point
V_f	volume of averaging domain filled with fluid
V_m	part of averaging domain visited by fluid during averaging time

V_p	volume of a particle
V_q	quadratic averaging domain
V_x	averaging domain with varying streamwise extent
V_{tot}	total volume of computational domain
V_x	averaging domain with varying streamwise extent
v_i	velocity vector of the fluid-particle interface
W	channel width components x_1, x_2, x_3
$\mathbf{x}$	coordinate vector with
$x_1 = x$	streamwise coordinate
$x_2 = y$	wall-normal coordinate
$x_3 = z$	spanwise coodinate
$\mathbf{x}_p$	coordinate vector of center of mass of a particle with components x_p, y_p, and z_p
y_w	artificial wall coordinate

Greek Symbols

α	slope angle of an experimental flume
Γ	boundary
γ	clipping function
Δ	difference operator
Δ_t	fluid time step
ϵ	model constant
ζ_n	surface distance of colliding particles
θ	scalar fluid quantity
λ_x, λ_z	typical period length in x- and z-direction
Λ	horizontal plane of an averaging window
μ	coeffiction of friction
μ_f	dynamic viscosity
ν_f	kinematic viscosity
ξ_x, ξ_y, ξ_z	transformed x-, $y-$, and z- coordinate
$\boldsymbol{\xi}_x, \boldsymbol{\xi}_y, \boldsymbol{\xi}_z$	local distance vector
π	non-dimensional quantity
ρ	density
ρ'	relative submerged density
$\boldsymbol{\tau}$	hydrodynamic stress tensor
τ	shear stress
τ_f	characteristic fluid time
τ_p	particle response time
τ_w	wall shear stress
$\tilde{\tau}_w$	modified wall shear stress
ϕ	porosity
ϕ_{AT}	space time porosity averaged over horizontal plane of domain
ϕ_T	temporal porosity

ϕ_V	spatial porosity
ϕ_{VT}	space-time porosity
ϕ_{Vm}	porosity of mobile particles
φ	particle quantity
ψ	void fraction
Ψ	integrated void fraction
Ψ_{in}	impact angle
Ω	particle control volume
$\boldsymbol{\omega}$	angular velocity vector with components ω_x, ω_y, and ω_z
Ω	computational domain

Subscripts

$aver$	averaging
bal	ballistic
$circ$	circumferential
$core$	center of mass coordinate of a particle cluster
$crit$	critical value
dry	dry collision
end	ending value
f	fluid property
fix	particles that do not move
$ideal$	ideal condition
in	before collision
$init$	initialisation
loc	locally averaged value
max	maximum
min	minimum
mob	particles that may move
n	normal
out	after collision
p, q	particle quantity
$proc$	processor
reg	linear regression analysis
rms	root mean square
sec	secondary current
sed	sediment
$start$	starting value
t	tangential
tot	total

Superscripts

$+$	value normalized by friction velocity or viscous length scale
(p)	*a posteriori* value
c	conditioned quantity
col	collision
cp	contact point
$crit$	critical value
lub	lubrication
i, j, k	space-discretized indices
m	moving
n	number of averaging time interval
r	resting
s	superficial average
p, q	particle quantity

Other symbols

$\overline{(.)}$	time-averaging operator
$\widetilde{\overline{(.)}}$	deviation of time-averaged value from its space-averaged value
$\langle . \rangle$	averaging operator
$\langle . \rangle_{x,y,z,t}$	averaging operator in x-, y-,and z-direction and time
$\lfloor . \rfloor = floor(.)$	largest previous natural number
$\lceil . \rceil = ceiling(.)$	smallest following natural number
$\lvert . \rvert$	absolute value
$(.)'$	deviation of an instantaneous value from a reference value
$\dot{(.)}$	derivative in time
$\mathcal{H}(.)$	Heaviside-function
$\mathcal{T}_0, \mathcal{T}_f,$	interval in time for intergration
$\mathcal{V}_0, \mathcal{V}_f,$	interval in space for intergration
$max(.)$	maximum value

Abbreviations

ACM	Adaptive Collision Model
ACTM	Adaptive Collision Time Model
ATFM	Adaptive Tagential Force Model
CDF	Cartesian distribution function
CFL	Courant-Friedrich-Levy
DOF	degree of freedom
DNS	Direct Numerical Simulation
IBM	Immersed Boundary Method
LM	lubrication model
LSM	Least Square Method
PRIME	Phase resolving simulation environment
RANS	Reynolds-Averaged Navier-Stokes
RPM	Repulsive Potential Model

1 Introduction

1.1 Background

Turbulent flows over mobile beds are ubiquitous features of many environmental and engineering systems. They largely determine systems performances in many processes in environmental as well as in process engineering, where the design of facilities such as bridge peers or sewerage and pipeline systems has to account for the nature of these flows. Therefore, a reliable prediction in terms of mass, momentum and energy transport has attracted significant attention, particularly in environmental engineering [160]. Typically, rivers carry a vast amount of sediment (Fig. 1.1). The transport modes of the sediment can be subdivided into two kinds [59]. Either the particles are small and light enough to be suspended by the turbulent nature of the channel flow [109], or with their size and density being large enough, they settle onto the rough sediment bed. The latter is referred to as the so-called bed-load mode, which is a a rolling, sliding or saltating motion [16, 60, 118].

Whenever heavy solid particles are transported by a carrier fluid, the prediction of the bed-load transport rate or the hydraulic resistance of the mobile bed, to mention but a few, becomes an important issue [139, 61]. The mechanisms driving the particle transport are very complex involving vortex structures and particle structures of disparate scales in space and time [37]. Scientist have tackled this problem since the early 20th century [138] as reviewed by [24], who also acknowledged the limited accuracy of common statistical approaches. A detailed understanding of the mechanisms that govern fluid–particle interaction is still lacking and raises the need for highly resolved data sets produced under controlled flow conditions. This, indeed, can be achieved by conducting Direct Numerical Simulations (DNS) of such phenomena, which offer access to the desired detailed data. This approach however requires an appropriate representation of the physical configuration targeted. One important issue in this respect is the use of a suitable particle collision model, which is capable to represent sediments with high volume fractions that are transported in the contact-dominated bed-load mode. Yet, another issue is the proper size of the computational domain and the duration of the averaging time to represent the spatial and temporal scales of the processes in question as well as the choice of proper statistical tools that are capable to answer the questions addressed.

1.2 Experimental efforts

Shields was among the first who formulated a framework for thresholds of incipient motion by an empirical correlation [141] The approach of Shields is based on a simplified balance of forces acting on bed particles considering the ratio of the destabilizing forces (drag, lift) to resistive forces (gravity, friction), today known as the Shields parameter. In fact, most of

Figure 1.1: *Estuary of the river Rhine into lake Constance (picture taken from [5]).*

the efforts to predict bed-load transport, such as the work by [103, 14, 15, 52], remain to be based on the 75-year old framework of Shields. The critical value of this parameter, which separates particle stability from entrainment by a turbulent flow, was found to be dependent on the particle Reynolds number. The Shields parameter is typically used for the assessment of both bed stability and bed-load transport, which is assumed to be proportional to the excessive bed shear stress, i.e. the difference between value of a given configuration and its critical value [122]. Over the years, the progress to describe and predict such processes, however, has been slow and this approach is still widely used in hydraulic applications [65]. This is due to lack of better models as reviewed by [24] as it is well known that these formulae introduce a high level of errors in estimates of bed stability and bed-load transport often exceeding 100% [18].

The particle transport mechanisms are complex and often involve highly organized structures such as multi-scale 2D- and 3D-bed forms moving with different speeds. At field scale, streamwise-oriented ridges were reported by Karcz [84] for small flow velocities and small mass loading. These structures are induced by secondary currents forming vortex tubes in and over the troughs [9]. The sediment pattern can be modified substantially by varying the sediment supply. Indeed, different types of particle structures were witnessed in experimental flumes for similar uniform flow conditions, when the mass loading was changed [47]. In this study, the relative density of the sediment was close to the critical threshold of incipient motion based on the condition of Shields [141]. Runs with low sediment supply produced ridges, confirming the observations reported above. For larger mass loadings on the other hand, spanwise oriented dune-like structures were observed. Due to the multi-disperse sediment used for these experiments, preferential transport changed the constitution of the bed load for different runs and exact distinction of the different physical mechanisms was not possible [127]. Experiments for different types of sediment and flow conditions but without additional sediment feeding were conducted by Shvidchenko & Pender [142]. Similar to the observations by Karcz [84], ridges were reported for moderate transport rates with a typical spanwise spacing of $2H$ and coherent structures in troughs with a streamwise extent of $12H$, where H is the flow depth of the channel [110].

Despite the observations cited above and other studies in the literature, a detailed understanding of the mechanisms responsible for the generation of theses structures is still lacking and further investigation of the fundamental processes is needed to develop tools for the description and modeling of such flows. This motivates new studies into mechanics of bed-load transport and its effects on the overall flow dynamics [38, 50, 46, 11].

1.3   Numerical efforts

Although experimental technologies for obtaining data sets of high resolution in both space and time are emerging [30], it will take some time before they are used for revealing physical mechanisms and testing theoretical predictions. Only recently sufficient computational resources have become available to conduct DNS of such phenomena on a reasonable scale. Therefore, DNS provide an alternative or complimentary approach to assess the data. For shear flows with dilute suspensions, a Lagrangian point-particle approach is suitable to simulate flows laden with suspended particles [8]. This type of simulation can account for particles smaller than the Kolmogorov scale and treats them as mass points. For turbulent channel flow, it was shown that particles tend to accumulate in the near-wall region with an uneven distribution, because particles deposited in low-speed streaks can be entrapped for long lapses of time [100]. Even though the point-particle approach can be further enhanced to four-way coupling of the two phases by modeling the feedback of the particles on the flow as well as effects from particle-particle collisions [10], it encounters limitations when dealing with high concentrations of sediments or with particle sizes larger than the Kolmogorov scale. The empirical correlations typically used in this framework were derived for undisturbed flow conditions in the first place and, therefore, fail to correctly describe the forces and torques of the particles for dense regimes [144].
Aware of these drawbacks, such simulations have been performed by Moreno & Bombardelli [106] with the focus on the importance of particle-particle collisions in sediment saltation. This is problematic since bed-load transport is characterized by high volume fractions of the disperse phase and very dense particle clusters close to the sediment bed [59]. The flow of the bed load hence is dominated by collisions and frictional forces. This underlines the need for fully-resolved simulations with four-way coupling of the flow and the disperse phase [17] and the necessity of an enhanced complexity of the numerical modeling of the collision process, a fact also supported by experiments [119]. Hence, a more sophisticated approach is needed to simulate such dense systems properly. This requirement is met by the Immersed Boundary Method (IBM), which has proven to be a valuable tool for conducting efficient simulations of multiphase flows that require full resolution and full coupling of the disperse phase and the fluid phase [83, 148, 88].
Computational studies of this kind on a large scale are still rare as the resolution requirements lead to very costly simulations of large domains over long time intervals. As a matter of fact, only a few years ago it was doubted that this was feasible at all at the required scale [17]. First simulations of horizontal, turbulent open channel flows, albeit marginally resolved, were conducted in [36] considering particles close to the suspended regime mimicking the experiments of [92], where particles are further away from the bed on average, so that collisions and contact are not as important. In [44], investigations were conducted for low particle Reynolds number with laminar flow conditions. Under turbulent conditions at low Reynolds number, particles with higher inertia were found to settle onto the bottom and to increase the intensity of turbulent fluctuations in this region [140]. Once having settled, particles arrange in typical patterns reflecting the equilibrium between erosion and deposition and the intricate interaction between fluid and particles. With low mass loading, the studies using a point-particle approach, such as [100], were confirmed by the particle-resolving simulations in [91] showing that particles accumulate in low-speed streaks. Simulations with higher particle Reynolds number on a coarse grid indicate that particles can be mobilized by extreme flow events even though the particle weight is larger than the critical threshold

of incipient motion [76].

In all of these studies, it was acknowledged, that particle-particle interaction by collision and contact is a crucial issue for the proper simulation of near-bed particle transport. Collision models accounting for friction have been derived in the framework of the Discrete Element Method (DEM). Such models were employed in simulations performed by Papista *et al.* [125], Osanloo *et al.* [121], Yergey *et al.* [162] and Duran *et al.* [49]. In Papista *et al.* [125], the shape of the particles was spatially resolved using a fictitious domain method. These authors, however, only considered two-dimensional configurations. Nevertheless, Papista *et al.* [125] concluded from their results that the choice of the collision model does not drastically affect the particle as well as the fluid motion. This is in contrast to the results of Moreno & Bombardelli [106] obtained with a point-particle approach. More sophisticated simulations were performed by Derksen [44] who investigated the incipient motion of spherical particles in a laminar shear flow near the critical Shields number using a lattice-Boltzmann method. A hard-sphere collision model was employed in combination with an explicit lubrication force for under-resolved viscous forces during the approach and rebound of colliding particles. The relevance of the collision model on the physical results and their sensitivity, however, was not investigated.

1.4   Research objectives and overview

The overall aim of this thesis is to carry out Direct Numerical Simulations of bed-load transport to analyze highly-resolved data sets in detail. This has do be done with a method that is capable to resolve the disperse phase properly and to display the relevant processes, which arise from particle-particle interaction. The method employed in the present thesis is outlined in Chap. 2. It is shown that the Immersed Boundary Method is the tool of choice to account for the relevant fluid-particle interaction as described by [17]. Chap. 3 is dedicated to the proper representation of particle-particle interaction, i.e. collision and contact. First, a review of the collision models described in literature is given and, subsequently, a systematic investigation of the impact of the collision models on the processes involved in bed-load transport are described. In particular, the classical model by Glowinski *et al.* [69] is compared to the recently proposed Adaptive Collision Model (ACM) [87], which unites several sub-models together with a temporal stretching of the phase of direct surface contact which is crucial for efficiency. Switching on and off different sub-models in the ACM and repeating the same simulation, or replacing the ACM with the model of [69] altogether, allows to access, which elements have to be present in a collision model to warrant physical realism. The question addressed here is whether the classical approach is sufficient or whether higher sophistication is required.

Unfortunately, large computational domains result in very costly simulations and an enormous amount of data that need to be convoluted into a handy set of parameters suitable for engineering applications, up-scaling, and physical interpretations. This requires a rigorous analytical framework for sediment motion and flow dynamics over mobile beds, which is currently under intensive development [112, 13]. In turn, these frameworks should provide a sound platform for studying the multifaceted mechanisms of sediment entrainment, transport, and deposition. A crucial issue is up-scaling and incorporation of grain-scale processes into large-scale models representing the entire bed-load or even the whole flow scale. Linking sediment mechanics with flow dynamics through coupled hydrodynamic equations rather than employing empirical correlations or intuition seems an appropriate strategy for this purpose. The development of these ideas in sediment transport research remains slow due to the lack of high resolution data in relation to both time and space.

This dilemma is addressed in Chap. 4, where time- and space-averaged hydrodynamic equations are applied for both flow regions within and above the fixed and mobile granular beds [115, 112] following an approach known as the Double-Averaging-Methodology (DAM) [116]. The DAM stems from the terrestrial canopy aerodynamics [131] and is described in detail, e.g., by [58]. During recent years it has also been successfully employed for studies of open-channel flows over fixed rough beds, examples of which are given in [117] and [104]. Among other findings, these studies highlight the importance of additional, dispersive fluid stresses, with their appearance being attributed to local heterogeneities of the time-averaged flow. The goal of Chap. 4 is to study, for the first time, the double-averaged momentum balance of the fluid phase in flows within and above mobile granular beds composed of monodisperse, cohesionless particles of a finite size using highly resolved DNS data.

In Chap. 5, a systematic study varying key-parameters of bed-load transport is carried out. The scenario considered is the particle-laden flow over a rough wall in a very large computational domain, which allows the disperse phase to develop bed forms with high fidelity, hence giving detailed insight into mechanisms of particle–fluid interaction. The characteristic numbers of the disperse phase considered here are similar to those of experiments in laboratory at medium Reynolds number [47, 142, 29]. The goal of the study is to reproduce patterns that are known from experimental evidence at a Reynolds number of the transitionally rough regime. Using the DAM-framework, the modification of turbulent fluctuations as well as the hydraulic resistance is assessed. Afterwards, this modification is explicitly linked to the time-averaged particle patterns and bed elevations investigated by suitable statistical tools.

While Chaps. 4 and 5 revolve around global or integral quantities observed for well developed flows, Chap. 6 is devoted to investigate particular flow events by means of conditional averaging. This allows a detailed investigation of key-mechanisms of singular events that can destabilize a developed particle order. One example is the investigation of coherent fluid structures that have the power to erode a single particle out off a closed layer of resting particles. These kinds of events are very rare, if the particle is heavier than its critical threshold of incipient motion, and must be detected by suitable conditions. Subsequently, conditional average of independent flow events in a local coordinate system with its origin at the center coordinate of the particle in question allows for detailed description of the typical fluid structure responsible for particle entrainment. Another example is the flow around moving particle clusters, which propagate with a significant velocity in streamwise direction, but are stable in time. To resolve the flow around these clusters, a transformed coordinate system is needed that moves along with the center coordinate of the particle clusters. This measure yields a stationary particle pattern such that stabilizing mechanisms like large-scale modifications of the flow field and destabilizing mechanisms such as the growth and decay of a large scale cluster can be studied in more detail.

Chap. 7 summarizes the main findings of Chaps. 3 - 6 and an outlook is given to employ the method in the future for further physical analysis and model development. The appendix contains technical issues about key-parameters of the given physical setup, scaling properties of the method employed for massively parallel applications, and the determination of initialization period to reach well developed conditions in the particle-laden channel flow.

2 Numerical method

2.1 Principal setup of an open-channel flow

The present work aims to contribute to the massive experimental effort carried out in the past to investigate turbulent flows over rough walls as well as bed-load transport of cohesionless particles [141, 14, 55, 47, 120, 29, 46]. All these experiments were conducted in an open channel under uniform flow conditions, i.e. the water depth does not vary in space and time. Moreover, measurements were carried out for developed flow conditions to exclude artificial effects stemming from the inlet conditions. The flow is driven by gravity with a fixed slope of the flume. To further simplify the problem, it is desirable to maximize the ratio of width of the flume W to water depth H, because side-walls induce secondary currents that may alter the flow conditions under consideration.

The scenario of a turbulent channel flow has also been a typical test case for various numerical studies [93, 107, 130, 62, 73]. In these references, the flow in a duct with smooth walls on the upper and lower boundary is considered. Unlike the experimental setup, the channel has no slope, but the flow is driven in streamwise direction (x−coordinate) by a pressure gradient. Hence, the flow becomes anisotropic because of the walls in y-direction referred to as the wall-normal coordinate. No side-walls are introduced, but periodic conditions are employed in the spanwise direction (z−coordinate) as detailed in Sec. 2.2 below.

In the present study, Direct Numerical Simulations (DNS) of an open-channel flow laden

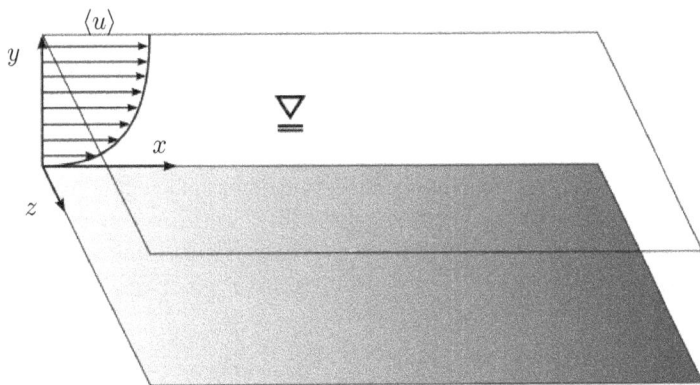

Figure 2.1: *Sketch of an open-channel flow with a flat plate on the bottom and an open surface at the top with stationary, uniform height as well as the the time-averaged velocity profile.*

with heavy particles are considered with a channel height H and a rigid-lid opposing no skin friction against the flow as indicated by the time-averaged velocity profile in the sketch shown in Fig. 2.1. The simulations were carried out with the code PRIME (*Phase resolving simulation environment*) developed at the *Institut für Strömungsmechanik* at TU Dresden, which is capable to perform simulations of multiphase flows using the Immersed Boundary Method (IBM). This chapter gives a detailed description of the code PRIME starting with the numerical procedure of the fluid solver including the boundary conditions and the generation of a channel flow in Sec. 2.2 followed by the coupling of the fluid phase and the particles in Sec. 2.3 and the models used for the representation of particle contact and collision.

2.2 Fluid solver

2.2.1 Governing equations

The fluid solver of the code PRIME is described in detail in [86] and used for the studies presented in this thesis. It solves the unsteady Navier-Stokes equations for an incompressible Newtonian fluid

$$\frac{\partial \mathbf{u}}{\partial t} + \nabla \cdot (\mathbf{uu}) = \frac{1}{\rho_f} \nabla \cdot \boldsymbol{\tau} + \mathbf{f} + \mathbf{f}_{IBM} \quad , \tag{2.1}$$

with $\boldsymbol{\tau}$ being the hydrodynamic stress tensor

$$\boldsymbol{\tau} = -p\,\mathbf{I} + \mu_f \left(\nabla \mathbf{u} + (\nabla \mathbf{u})^T \right) \quad , \tag{2.2}$$

and the continuity equation given by

$$\nabla \cdot \mathbf{u} = 0 \quad , \tag{2.3}$$

with $\mathbf{u} = (u, v, w)^T$ designating the velocity vector in Cartesian components, ρ_f fluid density, p pressure, μ_f the dynamic viscosity, $\mathbf{f}$ the volume force driving the mean flow, $\mathbf{f}_{IBM}$ an artificial volume force introduced by the IBM as described in Sec. 2.3 below, $\mathbf{I}$ the identity matrix, and t time.

2.2.2 Spatial discretization

The spatial discretization of Eq. (2.1) and (2.3) is performed by a second-order finite volume scheme on a fixed Cartesian grid with staggered arrangement of the fluid variables (Fig. 2.2a). Acknowledging the requirement to perform a true DNS, the grid spacing of the computational domain must be small enough to resolve the viscous scales of the flow [130]. In the present case of wall-bounded turbulence the flow becomes anisotropic in wall-normal direction. The smallest scales, hence, can be determined by

$$u_\tau = \sqrt{\frac{\tau_w}{\rho_f}} \quad , \quad l_\tau = \frac{\nu_f}{u_\tau} \tag{2.4}$$

with u_τ the friction velocity, τ_w the shear stress at the wall, $\nu_f = \mu_f / \rho_f$ the kinematic viscosity, and l_τ the viscous length scales [78, 62]. It was shown in [?], that this criterion is also applicable to turbulent flows over rough walls in the transitionally rough regime. Hence, the

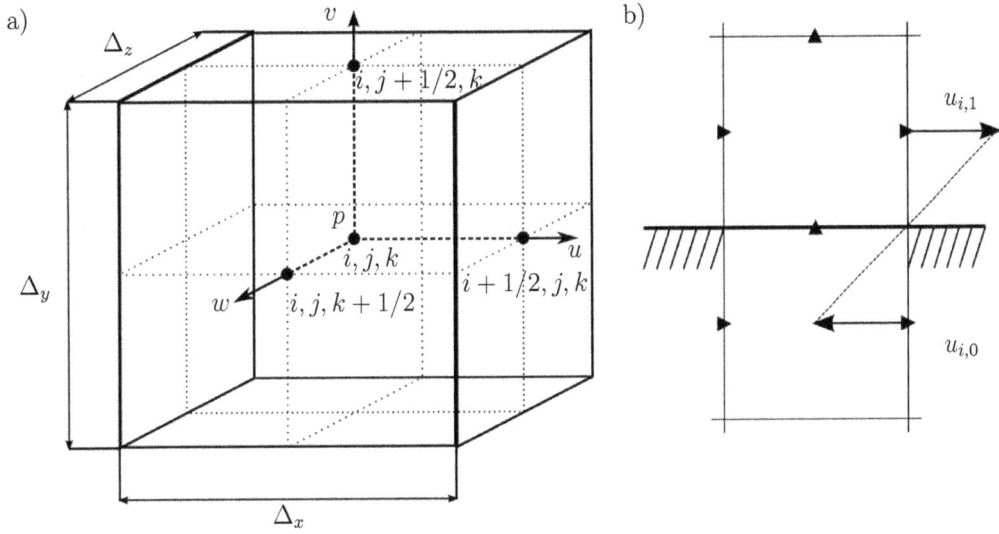

Figure 2.2: *Compational grid: a) cell with staggered arrangement of pressure and velocity component and b) linear interpolation for the realization of a no-slip condition.*

discretization in wall-normal direction Δ_y must satisfy $\Delta_y < l_\tau$. Due to reasons discussed in detail in Sec. 2.3, an Eulerian grid with isotropic cell size $\Delta_x = \Delta_y = \Delta_z = h$ was used to properly employ the IBM.

2.2.3 Time integration

The time integration is based on an explicit, low-storage Runge-Kutta three-step method of third order for the convective terms. This scheme comprises i) a fully explicit sub-step for the convective and the viscous terms yielding a preliminary non divergence-free velocity field, ii) the solution of the implicit Helmholtz-equation for the viscous terms resulting in an semi-implicit Crank-Nicolson scheme for the viscous terms altogether, iii) the Poisson equation that gives a pressure correction variable, and finally iv) the computation of the divergence-free velocity and pressure field based on the pressure correction. The explicit sub-step i) described above yields a restriction for the maximum of the time step $\Delta_{t,max}$ resulting in the well-known criterion of the Courant-Friedrich-Levy (CFL) condition

$$1 \geq \Delta_{t,max} \left(\frac{|u|}{\Delta_x} + \frac{|v|}{\Delta_y} + \frac{|w|}{\Delta_z} \right) \quad , \tag{2.5}$$

which is the ratio of the time step size to a characteristic convection time. This criterion applies to situations, where transport by convection is much more important than by diffusion.

2.2.4 Boundary conditions

The present setup consist of an open-channel flow that features a free-slip condition at the top wall and a no-slip condition at the bottom wall together with periodic boundary conditions in streamwise and spanwise directions. These conditions are realized by introducing a ghost-cell at the boundary Γ of the computational domain, where a desired velocity is imposed [56]. Whenever walls are introduced, this desired velocity can be either determined by a Dirichlet condition, i.e. the velocity of the fluid at the wall is equal to the velocity of the wall, or a Neumann condition imposing a velocity to obtain a desired gradient. The velocities at the ghost points are determined by extrapolating linearly from the first grid cell at the domain boundary through the desired value at Γ to the ghost cell (Fig. 2.2b). This procedure guarantees that

$$u_\Gamma = v_\Gamma = w_\Gamma = 0 \qquad (2.6)$$

is met for the no-slip condition, which is a homogeneous Dirichlet condition. The free-slip condition at the top wall requires

$$\frac{\partial u_\Gamma}{\partial y} = \frac{\partial w_\Gamma}{\partial y} = 0; \quad v_\Gamma = 0 \quad , \qquad (2.7)$$

which is a mixture of a homogeneous Neumann condition for u and w and a homogeneous Dirichlet condition for v. Note that the free-slip condition imposes a rigid lid allowing no fluid to leave the domain through the boundary. With H being constant, this configuration is similar to stationary and uniform flow conditions in laboratory setups.

The periodic boundary conditions are realized by equalizing the corresponding values at the beginning and the ending of the computational domain in the given direction

$$\mathbf{u}(0, y, z) = \mathbf{u}(L_x, y, z); \quad \mathbf{u}(x, y, 0) = \mathbf{u}(x, y, L_z) \qquad (2.8)$$

with L_x the streamwise and L_z the spanwise extent of the computational domain. This approach requires a size of the computational domain large enough to represent the largest energy-containing motions, because fluid information can be passed in downstream and upstream direction throughout the domain, if sub-critical conditions prevail [130]. A sufficient domain size inhibits the interaction of large-scale coherent fluid structures, such as the well-known hairpin vortices [6], with itself over the periodic boundaries. A decorrelated fluid field would satisfy the requirement described above, which can be quantified by the two-point correlation discussed in Sec. 5.3 below.

2.2.5   Driving the channel flow

Looking in streamwise direction, the simulation of a turbulent channel is similar to recirculating flumes [127], but instead of going through a pumping system to compensate the slope of the flume, the volume force $\mathbf{f} = (f_x(t), 0, 0)^T$ is introduced as a source term on the right-hand side of Eq. (2.1) driving the mean flow through a channel with zero slope. This volume force is comparable to a pressure gradient. The volume force is constant in space, but adjusted dynamically in time to maintain a constant fluid mass flux in the channel. Note that with this method, the bulk velocity U_b is fairly kept constant in time and the friction velocity u_τ is a result of the simulation. In doing so, the statistically stationary state is reached faster than by imposing a constant volume force [62]. The driving force is determined by a PI-controller

$$f_x(t) = K_p \, \Delta U_b(t) + K_i \int \Delta U_b(t) \, \mathrm{d}\, t \qquad (2.9)$$

where the quantity $\Delta U_b(t) = U_{b,a}(t) - U_b$ is the difference between the actual bulk velocity $U_{b,a}$ and the desired bulk velocity U_b. The constants in (2.9) are chosen empirically to be $K_p = 24.0$ and $K_i = 30$ such that the maximum deviation from the desired velocity ΔU_b is 0.1% [86].

2.3 Immersed boundary method

Direct Numerical Simulations (DNS) are a powerful tool for providing high-resolution data of bed-load transport. In cases when particles are larger than the dissipative Kolmogorov scale, point-particle approaches cannot be used without additional modeling that involves well-known uncertainties. For example, such point-particle approaches particularly suffer from the required empirical correlations for drag and lift if many particles are close together or if particles are colliding [144]. Bed-load transport is characterized by high volume fractions of the solid phase and very dense particle clusters close to the sediment bed. This bed-load feature underlines the need for fully-resolved simulations with four-way coupling of the disperse phase [17]. The method of choice for this kind of problem is, therefore, the Immersed Boundary Method to represent solid particles numerically. Parts of the following description of the IBM are integrated in a manuscript published in the *Journal of Multiphase Flow* [89].

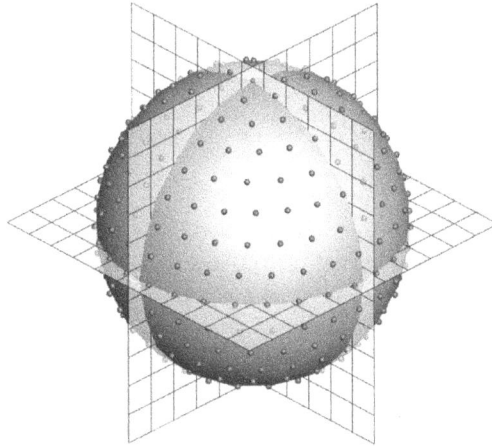

Figure 2.3: *Particle surface with 315 discrete marker points and three slices of the Cartesian background grid.*

2.3.1 Representation of the particle-fluid interface

First types of the IBM for moving particles were presented in [83] and [82]. The method of the present studies is based on the scheme proposed by [148]. The fluid-particle interface is represented by discrete Lagrangian marker points. The coupling of the continuous and the disperse phase is realized by the additional source term $\mathbf{f}_{IBM}$ in the Navier-Stokes equation (2.1) in the vicinity of the particle. These forces are computed such that they impose a no-slip condition for the fluid at the particle interface. This force is determined by a direct-forcing approach [105, 54]. For this matter, the transfer between fixed Eulerian grid cells and the Lagrangian marker points is performed by interpolation and spreading operations via regularized Dirac delta functions [128]. In the present implementation, the three-point function of [135] was used. The number of marker points needed can be estimated by the assumption given in [148] that the marker points must cover a spherical shell of thickness h around the particle with each marker point representing a volume of $\Delta V_l = h^3$ of this shell. This assumption ultimately results in

$$N_l \approx \frac{\pi}{3}\left(12\frac{D^2}{4\,h}+1\right) \tag{2.10}$$

with N_l the number of Lagrangian marker points, if an Eulerian grid with an isotropic cell size of $\Delta_x = \Delta_y = \Delta_z$ is used. This is indeed the case for the present study. The Lagrangian marker points were distributed evenly over the particle surface by the algorithm of [97] to match this requirement of the discretization (Fig. 2.3).

2.3.2   Computation of particle motion

The motion of each individual spherical particle is calculated by solving an ordinary differential equation for its translational velocity $\mathbf{u}_p = (u_p, v_p, w_p)^T$

$$m_p \frac{\mathrm{d}\mathbf{u}_p}{\mathrm{d}t} = \oint_{\Gamma_p} \boldsymbol{\tau}\cdot\mathbf{n}\,\mathrm{d}s + V_p\left(\rho_p - \rho_f\right)\mathbf{g} + \mathbf{F}_p \tag{2.11a}$$

and for its angular velocity $\boldsymbol{\omega}_p = (\omega_{p,x}, \omega_{p,y}, \omega_{p,z})^T$,

$$I_p \frac{\mathrm{d}\boldsymbol{\omega}_p}{\mathrm{d}t} = \oint_{\Gamma_p} \mathbf{r}\times(\boldsymbol{\tau}\cdot\mathbf{n})\,\mathrm{d}s + \mathbf{M}_p \quad . \tag{2.11b}$$

Here, m_p is the particle mass, V_p the particle volume, g the gravitational acceleration, $I_p = 8\pi\rho_p R_p^5/15$ the moment of inertia, ρ_p is the particle density, and R_p the particle radius. The vector $\mathbf{n}$ is the outward-pointing normal on the interface Γ_p, and $\mathbf{r} = \mathbf{x} - \mathbf{x}_p$ is the position vector of the surface point with respect to the center of mass $\mathbf{x}_p$ of the particle. The term $\mathbf{F}_p$ denotes the forces resulting from particle-particle interaction and $\mathbf{M}_p$ is the moments generated by these interactions, which are described in detail in Chap. 3 below. Note that the first term on the right-hand side of (2.11) and (2.11b) representing the stresses exerted on the particle by the fluid is not computed in a straightforward manner with the present method, but was substituted by the expressions

$$\rho_f \oint_{\Gamma} \boldsymbol{\tau}\cdot\mathbf{n}\ \mathrm{d}s = \frac{\mathrm{d}}{\mathrm{d}t}\int_{\Omega_p} \rho_f\mathbf{u}\,\mathrm{d}V - \rho_f\int_{\Omega_p}\mathbf{f}_{IBM}\,\mathrm{d}V \tag{2.12a}$$

for the momentum balance of the translational velocity and

$$\rho_f \oint_{\Gamma} \mathbf{r}\times(\boldsymbol{\tau}\cdot\mathbf{n})\,\mathrm{d}s = \frac{\mathrm{d}}{\mathrm{d}t}\int_{\Omega_p}\rho_f\mathbf{r}\times\mathbf{u}\,\mathrm{d}V - \int_{\Omega_p}\rho_f\mathbf{r}\times\mathbf{f}_{IBM}\,\mathrm{d}V \tag{2.12b}$$

for the angular momentum with Ω_p a control volume equal to the particle volume. The time integration of (2.11a) and (2.11b) is based on the same Runge-Kutta scheme already described in Sec. 2.2 for the fluid phase.

2.3.3 Enhanced IBM

The computational method presented in this section is based on the enhanced IBM described by [88]. Unlike [148], no assumptions were made to compute the right-hand side of (2.12a) and (2.12b), but both terms, the change of momentum of the fluid enclosed in the particle and the total force introduced by the IBM acting on the particle, were directly computed by the enhancements of [88]. Using this method, the stability range of the IBM is significantly increased for particle density ratios ρ_p/ρ_f from 1.2 to 0.3. Moreover, the consistency of the spreading of the forces from the Lagrangian marker points to the Eulerian grid was improved for particles that are in direct contact by excluding all surface markers from the computation of forces of solid bodies, whose stencil overlap with the stencil of a collision partner. This issue becomes of particular importance when dealing with frequent particle contact and collisions as described in Sec. 3.2.

3 Choice of collision model

3.1 Introduction

Typically, particle motion in bed-load mode is dominated by collisions and particle contact. The studies carried out so far share the issue of a proper representation of collisional processes [36, 125, 44, 140, 91, 76]. These collisional processes between two particles, either both moving or one fixed and another moving, are key mechanisms of momentum exchanges and sinks. Their adequate modelling is therefore of major interest for the understanding of this type of flow as well as for its successful simulation. Indeed, in all the references cited above, the general structure of the modelling approach remains the same. Papista *et al.* [125], therefore concluded from their results that the choice of collision model does not drastically affect the particle as well as the fluid motion. These authors, however, only considered two-dimensional configurations. This is also in contrast to the results of Moreno & Bombardelli [106] obtained with a point-particle approach. The present chapter addresses this issue, which was a joint work contributing also to the dissertation of Tobias Kempe [86], who developed the sophisticated Adaptive Collision Model (ACM) for single and multiple particle collisions. In the present thesis, a comparative analysis of meaningful statistical quantities is presented to investigate the impact of different collision models on sediment transport. Most of the results shown in the following sections are incorporated in a manuscript presented first in 2011 to the *7th International Symposium on Turbulence and Shear Flow Phenomena* (TSFP 7) [153]. The study was eventually peer-reviewed and published in the *Journal of Multiphase Flow* [89].

The physical situation of bed-load transport in a three-dimensional turbulent open channel flow over a rough wall is considered here, with parameters chosen such that the sediment is close to incipient motion. The Shields number is varied together with other parameters to address the effect of different regimes on the particle behavior. All these cases involve turbulent flow, motivated by the observation of Yalin & da Silva [160] that sediment forming natural bed forms is related to turbulent conditions. On the other hand, the Reynolds number has to be moderate for reasons of feasibility of the simulations. Nevertheless, the data being generated provide valuable new and detailed information on collision modeling and the behavior of bed-load sediment. The classical regime diagram of Clift *et al.* [39] features, at the upper end of the mass loading coordinate collision-dominated and contact-dominated flows. These in fact are the regimes covered by bed-load transport of sediment. Due to the high local mass loading, the collision model is required to work with multiple simultaneous collisions. This step is accomplished in the present paper by demonstrating that in fact the Adaptive Collision Model (ACM) proposed by Kempe & Fröhlich [88] proved to be very reliable for the test cases involving the contact-dominated transport a single and a multitude of mobile particles.

The chapter is organized as follows. Sec. 3.2 presents a detailed description of the collision models reported in literature and examples for the computation of the normal forces are given. Section 3.3 and 3.4 first present the specification of the physical configuration considered and then provide results for a single particle being transported over a rough bed. Here, only collisions between moving and fixed particles need to be modeled. The following Section 3.5 reports on simulations with many mobile particles, where collisions between mobile particles and between mobile and fixed particles occur.

3.2   Collision modelling

3.2.1   Structure of collision models

One of the major advantages of the IBM is the direct compuation of long-range interactions between the particles. Only short-range interactions as well as collisions need to be modelled. The total force $\mathbf{F}_p$ acting on a particle p during the collision process may be decomposed as

$$\mathbf{F}_p = \sum_{q,\,q\neq p}^{N_p} \left(\mathbf{F}_{n,pq}^{lub} + \mathbf{F}_{n,pq}^{col} + \mathbf{F}_{t,pq}^{col} \right) \quad , \tag{3.1}$$

where $\mathbf{F}_n^{col}$ is the normal and $\mathbf{F}_t^{col}$ the tangential force during the interface contact, and $\mathbf{F}^{lub}$ is the modelled lubrication force during approach and rebound. The torque $\mathbf{M}_p$ on a spherical particle p generated by the tangential contact forces is

$$\mathbf{M}_p = \sum_{q,\,q\neq p}^{N_p} R_p\, \mathbf{n}_{pq} \times \left(\mathbf{F}_{n,pq}^{lub} + \mathbf{F}_{t,pq}^{col} \right) \tag{3.2}$$

with $\mathbf{n}_{pq}$ the unit vector pointing from $\mathbf{x}_p$ to $\mathbf{x}_q$ and R_p the particle radius.
Collision modelling now amounts to providing suitable expressions for the normal and tangential forces introduced in (3.1) and (3.2).

3.2.2   Hard-sphere model

The idea of an hard-sphere model is to represent the exact solution of a particle trajectory (Fig. 3.1) by imposing

$$u_{p,out} = -e\, u_{p,in} \tag{3.3}$$

with $u_{p,in}$ the absolute particle velocity before and $u_{p,out}$ the absolute velocity after the collision. The restitution coefficient e of the mobile particles hereby describes the damping during collision. This implies that the collision process is non-elastic and not resolved in time, i.e. the contact time is assumed to be infinitesimally small. This modelling strategy was developed for point-particles approaches in the first place giving satisfactory results for simulations of dilute regimes [161]. The hard-sphere model was also applied in [44] for fully-resolved simulations of a granular bed sheared by a laminar flow. It was shown, however, in [87] that this approach fails to properly describe the exact solution (solid line in Fig. 3.1), because with fully-resolved simulations the fluid can not follow the discontinuous velocity change of the particle. It was, hence, found to be not suitable for the phase-resolved simulations and, therefore, not accounted for in the analysis presented in the following.

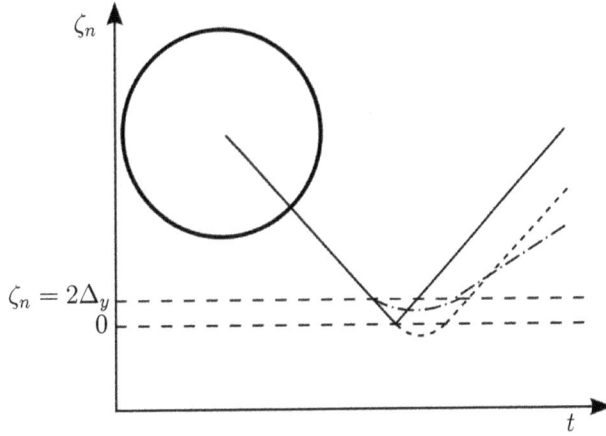

Figure 3.1: *Schematic trajectory of a particle impacting on a wall: exact solution (solid line), ACM (dashed), and RPM (dashed dot).*

3.2.3 Repulsive potential model

With the Repulsive Potential Model (RPM), the collisions are elastic preserving the total kinetic energy of the particles. Moreover, the full decelerating and accelerating motion of the particle is resolved by stretching the collision process in time. The normal forces introduced by the RPM reads [68]

$$\mathbf{F}_{n,pq}^{col} = \frac{\mathbf{x}_p - \mathbf{x}_q}{\epsilon} \left(\max\left\{0, -\left(\zeta_{n,pq} - S\right)\right\}\right)^2 \quad , \tag{3.4}$$

where ϵ is a model constant depending on the problem considered, $\zeta_{n,pq} = |\mathbf{x}_q - \mathbf{x}_p| - (R_p + R_q)$ is the distance of the colliding surfaces and S the range of the repulsive force. This approach has been standard in literature as it was used for the simulations presented in [149, 36, 125, 44, 66, 140, 91]. Here, the collision time becomes a function of the material stiffness ϵ, which is to be chosen *a priori*. If collisions of different physical types, e.g. a resting and a moving particle with a high impact velocity or two resting particles, occur during the course of a simulation, the choice of ϵ becomes delicate, because it must be calibrated for the strongest impacts possible, i.e. the collision with the shortest contact time, during the course of the simulation. While this type of collisions determines the stiffness, collisions of particles with a low relative velocity may become quite long, dampening the process in a unphysical manner as sketched by the dash-dot line in Fig. 3.1. Hence, the model is capable of representing sedimenting particles in a vertical channel like in [149], but the larger the range of relative velocities is during the course of a simulation, the larger the errors may become for the representation of collisions [87]. This is especially of concern for simulations of horizontal channels laden with heavy particles like [140]. Moreover, to avoid unresolved fluid in the gap between two colliding particles, $S = 2h$ is usally chosen. Note that if particles are resolved with $D/h \approx 10$ as it was done in [36, 91], the volume of a particle is significantly expanded by this artificial range of the repulsive force.

3.2.4 Adaptive collision model

The adaptive collision modell was first proposed by [87] with the goal to introduce a strategy for particle contact, which remedies the drawbacks discussed in Sec. 3.2.3. The motivation was to minimize the number parameters that have to be chosen *a priori*. This approach comprises the Adaptive Collision Time Model (ACTM) to calculate normal forces, the Adaptive Tangential Force Model (ATFM) to model tangential forces and a lubrication model to account for the effects of unresolved fluid in between the gap of two approaching particles. This approach has proven to be very suitable for simulations of particle-resolving simulations, especially if particles are transported in bed-load mode, i.e. with very frequent collisions [89].

Adaptive collision time model

Normal forces are computed using a generalization of contact model of Hertz [72]. Hence, the Adaptive Collision Time Model (ACTM) reads

$$\mathbf{F}_{n,pq}^{col} = -\left(k_n \, |\zeta_{n,pq}|^{3/2} + d_n \, g_{n,pq}\right) \mathbf{n}_{pq} \tag{3.5a}$$

$$m_p \frac{\mathrm{d}^2\zeta_n}{\mathrm{d}t^2} + d_n \frac{\mathrm{d}\zeta_n}{\mathrm{d}t} + k_n \zeta^{3/2} = 0 \quad , \tag{3.5b}$$

where k_n is the material stiffness, d_n is the damping coefficient, $g_n = -\,\mathrm{d}\zeta_{n,pq}/\mathrm{d}t$ is the normal part of the relative surface velocity at the closest point, and ζ_n is the normal distance between the surfaces of the colliding particles. In order to avoid excessive time step reduction during collision, which would be required with realistic material parameters, the duration of a direct contact is stretched in time treating the interval with $\zeta_n < 0$ as a dry collision. With the duration of the collision set to $10\Delta_t$, where Δ_t is the time step of the fluid solver, the ACTM determines k_n and d_n such that the damping and restitution coefficients are correctly represented allowing avoidance of the excessive time step reduction during collisions. The restitution coefficient of the particles describing the amount of damping during collisions was set to $e = -u_{p,out}/u_{p,in} = 0.97$. This value corresponds to the one of glass beads and, hence, is close to the one of sand grains [80].

Adaptive tangential force model

The tangential forces were computed with the Adaptive Tangential Force Model (ATFM), another sub-model of the ACM. Here, the criterion to distinguish between rolling and sliding motion is the local angle of impact

$$\Psi_{in} = \frac{\mathbf{g}_{t,pq}^{cp}}{\mathbf{g}_{pq} \cdot \mathbf{n}_{pq}} \quad , \tag{3.6}$$

where $\mathbf{g}_{pq} = \mathbf{u}_p - \mathbf{u}_q$ is the relative velocity of the particle centers and $\mathbf{g}_{t,pq}^{cp}$ the tangential part of the relative surface velocity at the contact point $\mathbf{x}_{pq}^{cp}$. If the impact angle is smaller than a critical value, in this case $\Psi_{in}^{crit} = 0.95$ as proposed by [87] for glass spheres, a rolling contact is assumed. When the two particles p and q are in rolling contact, the relative surface velocity g_t at the contact point vanishes at each instant in time. The model for this process is based on the idea of applying a tangential contact force in such a way that this condition is imposed exactly in time-discrete form by computing the desired force $\mathbf{F}_{t,pq}^{col}$ at each sub-step of the Runge–Kutta integration scheme. For sliding motion, the collision force in tangential direction is given by the friction law of Coulomb

$$\mathbf{F}_t^{col} = -\mu\,|\mathbf{F}_n|\,\mathbf{t}_{pq}^{cp} \quad ,$$
(3.7)

where the tangential coefficient of friction $\mu = 0.15$ is a parameter of the material, chosen according to the experimental data of [80], and $\mathbf{t}_{pq}$ is the unit vector collinear with the tangential force vector. This model constitutes a substantial improvement over previous approaches, since exact rolling is guaranteed by this approach.

Lubrication model

When the distance between the surfaces of two approaching particles becomes small, the fluid is squeezed out of the gap. Viscous forces are very important in this phase and can lead to significant dissipation. The ACM therefore includes a subgrid-scale model accounting for the viscous film between the particles when it is too thin to be resolved by the Eulerian grid to calculate the missing part of the hydrodynamic force using analytic expressions derived from Stokes flow approximations. This so-called lubrication model (LM) corrects the viscous and pressure forces acting upon the particles by explicit correlations used to determine the resulting force between the particles. The same procedure is applied for particles rebounding after the collision. The lubrication force is dissipative, since it is always directed opposite to the relative velocity. The lubrication corrections for particle-particle interactions employed in the ACM are based on the result of Cox and Brenner [42] and read

$$\mathbf{F}_{n,pq}^{lub} = \begin{cases} 0, & 2\Delta_x < \zeta_{n,pq} \\ -\frac{6\,\pi\,\nu_f \rho_f\,g_{n,pq}}{|\zeta_{n,pq}|}\left(\frac{R_p\,R_q}{R_p+R_q}\right)^2 \mathbf{n}_{pq}, & \zeta_{min}^{lub} \le \zeta_{n,pq} \le 2\Delta_x \\ 0, & \zeta_{n,pq} < \zeta_{min}^{lub} \end{cases}$$
(3.8)

Note that a similar expression derived by Nguyen & Ladd [111] was applied in the study of [44]. A cut-off distance ζ_{min}^{lub} is used to prevent the lubrication force from reaching its singularity at zero normal distance and, from a physical point of view, to account for the natural surface roughness. All computations below were carried out with $\zeta_{min}^{lub}\,/\,D = 10^{-4}$ with D the diameter of a particle.

Sustained contact

In the literature on multiphase flows an often employed distinction is the one between collision-dominated flows and contact-dominated flows [43]. Both situations are relevant for bed-load transport of sediment and can even occur in the same flow at different instants in space and time. In the present text, the term "contact" is primarily used to designate the short time interval during a collision, when the particle surfaces actually touch. To make the distinction clear, the term "sustained contact" is used in the following for the situation when a particle is resting on a wall or on top of other particles. The transition between both situations is characteristic for sediment mobilization and deposition, so that the model must be able to adequately account for this transition. The ACTM presented in [87] indeed is capable of doing so and is used in the present study without modification. Disregarding

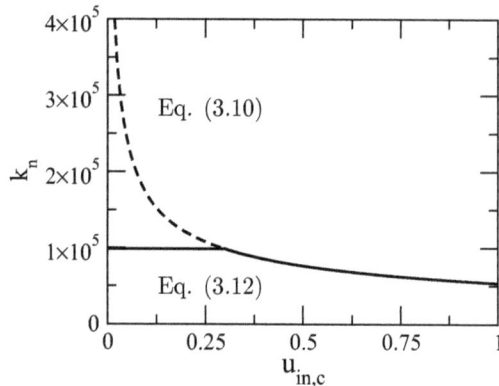

Figure 3.2: *Illustration of the limiter for sustained contact, here exemplary with a critical impact velocity of $u_{in,c}^{crit} = 0.3$. Below this value the stiffness k_n is kept constant according to Eq. (3.12). Without the limiter k_n in Eq. (3.10) is infinitely large for $u_{in,c} \to 0$.*

the motion of surrounding fluid, the force generated by the collision model and the gravity force must be in equilibrium. For steady state, the Hertzian contact model (3.5a) reduces to $F_{n,pq}^{col} = k_n \, |\zeta_{n,pq}|^{3/2}$, so that

$$k_n = \frac{m_p \, g}{|\zeta_{n,pq}|^{3/2}} \quad . \tag{3.9}$$

With realistic parameters, k_n usually is very large (e.g. $k_n = 2.648 \times 10^9$ Nm^{-1} for glass beads with $D = 3$ mm), and $\zeta_{n,pq}$ is extremely small. Large values of k_n can generate instabilities in the solution of the particle equation of motion. Furthermore, it does not blend with the values of k_n determined to obtain the modified collision time of the ACTM, $T_c = 10 \, \Delta_t$.

Consider a normal particle-wall collision using the ACTM with $u_{in,c}$ being the velocity of the particle just before touching the surface of the wall. Eq. (3.5a) with $d_n = 0$ yields the relation

$$k_n = 18.578 \frac{m_p}{\sqrt{T_c^5 \, u_{in,c}}} \quad . \tag{3.10}$$

Sustained contact is related to $u_{in,c} \to 0$ and infinite collision time. If, however, $T_c = 10\Delta_t$ is used as a fixed parameter in the simulation, this may lead to contradiction. The remedy proposed in [87] is to limit k_n from above by using the unmodified value $T_c = 10\Delta_t$ and a lower limit of $u_{in,c}$. The latter value $u_{in,c}^{crit}$ is found by observing that in a viscous flow no particle rebound occurs below a certain value of the Stokes number. Setting the Stokes number formed with $u_{in,c}^{crit}$ equal to 1 and setting the corresponding velocity to be the critical value yields

$$u_{in,c}^{crit} = \frac{9}{2} \frac{\rho_f \, \nu_f}{\rho_p \, R_p} \quad . \tag{3.11}$$

The final form of (3.11) used in the simulation hence is

$$k_n = 18.578 \frac{m_p}{\sqrt{T_c^5 \max\left\{u_{in,c}, u_{in,c}^{crit}\right\}}} \quad . \tag{3.12}$$

with T_c always given by the same multiple of Δ_t. This is illustrated in Fig. 3.2 Sustained particle-particle instead of particle-wall contact is treated in the same way, only the scale factor in (3.10) is different [87]. For $u_{in,c} > u_{in,c}^{crit}$ the duration of surface contact is T_c, as

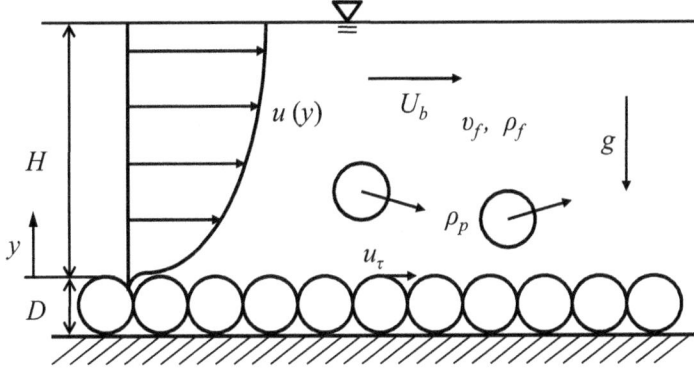

Figure 3.3: *Sketch of the flow above a rough wall in an open channel with transport of mobile particles. Definition of physical parameters describing the problem.*

conceived. For $u_{in,c} < u_{in,c}^{crit}$, T_c becomes just a parameter and the time of contact exceeds T_c. This way of limiting k_n via a bound imposed on the value of $u_{in,c}$ used in the formula for k_n allows to leave T_c untouched and furthermore to restrict the value of k_n to just the amount which is needed. Setting the damping coefficient to zero in these considerations is justified by the fact that damping has only little or no influence in case of small particle velocities $u_{in,c}$ and sustained contact.

3.3   Compuational Setup

3.3.1   Dimensionless parameters

The numerical setup used in the present chapter is an open channel flow with periodic boundary conditions in streamwise and spanwise direction. Spherical particles are introduced in the flow applying the same periodic conditions as illustrated in the sketch of Fig. 2.1. The bottom of the channel is constituted by one or three layers of fixed hexagonally packed particles, which have the same size and material properties as the mobile particles. An equidistant Eulerian grid with equal step size $\Delta_x = \Delta_y = \Delta_z$ in all directions is employed. The flow is driven by a pressure gradient, which is adjusted instantaneously in time such as to maintain a constant flow rate. The free-slip condition (2.7) is used at the top boundary which is common practice for this type of simulation [98, 162, 90].

As illustrated in Fig. 3.3, 10 physical parameters describe the present problem. These are the free height above the sediment H, the kinematic viscosity of the fluid ν_f, the fluid density ρ_f, and the characteristic velocity of the flow, i.e. the bulk velocity U_b in terms of the channel flow, and wall shear stress τ_w, which can also be linked to the characteristic friction velocity $u_\tau = \sqrt{\tau_w / \rho_f}$ for near-wall considerations. Furthermore, $V_{tot} = L_x L_y L_z$ is the total volume of the computational domain and the volume filled with fluid is $V_f = V_{tot} - N_p V_p$, which gives a volumetric measure to describe the density of the particle-laden flow with N_p the total number of particles introduced in the channel. In addition, the submerged particle density $\rho_p - \rho_f$, the diameter of the particles D, and the gravity g describes the particle properties. The present dimensional analysis extents the considerations of Ettema [53], Papista *et al.* [125], and Kempe *et al.* [89] by the wall shear stress and the volume fraction, because different configurations of particle arrangements can lead to different roughness elements

Re_b	L_x/D	L_z/D	H/D	Re_τ	D^+	D/h	h^+
4449	72	36	12	263	22	14	1.57

Table 3.1: *Key parameters of Setup $\mathcal{S}$ used to analyze single particle motion over a rough bed.*

Regime	$|\mathbf{g}|\,H\,/\,U_b^2$	$(\rho_p - \rho_f)\,/\,\rho_f$	Fr	Sh	Sh_{crit}
SA	9.81	0.25	2.45	0.0171	0.034
SB	9.81	0.75	7.35	0.0057	0.034
SC	9.81	1.25	12.26	0.0034	0.034

Table 3.2: *Dimensionless parameters characterizing the three regimes in the study with a single mobile particle.*

enhancing or diminishing hydraulic resistance. This effect was not included in the references cited above, but it has been acknowledged as one of the key parameters of sediment transport (cf. e.g. [47, 26]).

In accordance with the Buckingham Π-theorem (Sec. A), 7 characteristic numbers can be derived from the 10 physical parameters, which fully describe the setup:

the Froude number

$$Fr = \frac{U_b}{\sqrt{gH}} \quad , \tag{3.13a}$$

the Shields parameter

$$Sh = \frac{\tau_w}{(\rho_p - \rho_f)gD} \quad , \tag{3.13b}$$

the bulk Reynolds number

$$Re_b = \frac{U_b H}{\nu_f} \quad , \tag{3.13c}$$

the particle Reynolds number

$$D^+ = \frac{u_\tau D}{\nu_f} \quad , \tag{3.13d}$$

the relative volume of the domain $\frac{V_{tot}}{D^3}$, the relative submergence H/D, and the total porosity V_f/V_{tot}. The Shields parameter was derived in the framework of an experimental analysis of incipient motion [141] to quantify the mobility of a multidisperse sediment on the basis of a statistical approach. It compares the average frictional force to the gravitational forces inhibiting the mobilisation of a particle embedded in a sediment packing. Based on these experimental investigations, a critical Shields number, Sh_{crit}, is usually determined and used as an indicator for incipient sediment motion. Since only a few exposed particles are placed on top of the fixed bed in the present configuration, the Shields criterion is not directly applicable. For lack of a better quantity, it is used here nevertheless as a rough indicator for the mobility of the particles.

3.3.2   Configuration with a single mobile particle

The goal of the first setup, Setup $\mathcal{S}$, is to study the effects of the collision model on the trajectory of a single particle placed on top of a hexagonally packed sediment bed. The situation is sketched in Fig. 3.4 with the arrow pointing to the exposed mobile particle.

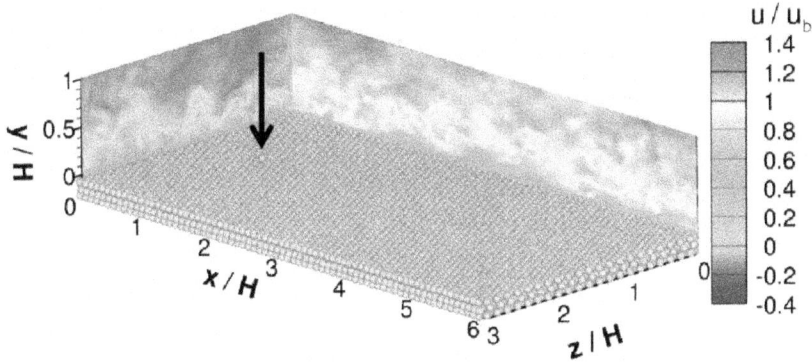

Figure 3.4: *Setup S according to Tab. 3.1 used to study the effect of collision modeling on single particle trajectories. Contour plots on the faces of the computational domain represent the instantaneous streamwise velocity component. The arrow points to the single exposed particle at its initial location.*

In a preliminary simulation the flow was initialized and advanced in time until a statistically stationary turbulent state was reached. Then, the exposed particle was released and mobilized by the turbulent fluid. Statistics of the particle trajectory and the particle velocities, both translational and angular, were recorded over a large period in time, more than 660 bulk units H/U_b. The Reynolds number of this setup was chosen to be rather high in order to simulate a wide range of possible states of motion of the particle, like creeping and jumping. The key parameters for this study and the various regimes in terms of the characteristic numbers are summarized in Tabs. 3.1 and 3.2.

As detailed above, a collision model can, or can not, account for different physical mechanisms, such as lubrication, normal contact forces and tangential contact forces. To elucidate the role of these components and determine the required level of modeling, several runs were conducted with different components being activated. To account for the normal forces, the RPM (3.4) and the ACTM (3.5a) are used in this study. Further investigations focus on the effect of tangential forces accounted for by the ATFM (3.7). The lubrication model (3.8) is switched on in all simulations with the ACTM since it is required to accurately predict the motion of the particle in the vicinity of the sediment bed. The ACTM without the LM yields unphysical results as demonstrated in [87]. Recall, that the ACM, as defined in Section 3.2.4 above, comprises the joint use of the LM, the ACTM and the ATFM.

3.3.3 Configuration with multiple mobile particles

To investigate the impact of the collision model in complex situations of bed-load transport, situations with a large set of mobile particles are investigated. The rough surface now is modeled by a hexagonally packed sediment bed consisting of a single layer with $N_{p,fix}$ fixed spherical particles. In addition, a number of $N_{p,mob} = 500$ mobile particles are placed on top of the fixed bed. This configuration is called Setup $\mathcal{M}$ in the following and is sketched in Fig. 3.5. The number of particles is relatively small so that the considered situation falls into the regime of small mass loading according to Dietrich *et al.* [47]. The domain size and the spatial resolution is the same as for Setup S. The bulk Reynolds number is $Re_b = 3010$. Two different regimes, summarized in Tab. 3.3, are considered in the present study. In

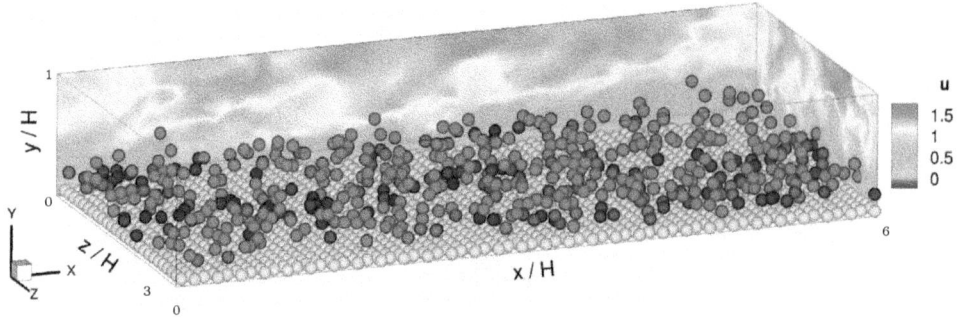

Figure 3.5: *Setup M used for the investigations of sediment transport with a fixed sediment bed (grey colour) and $N_{p,free} = 500$ mobile particles colored according to their velocity. Snapshot of the particle distribution and the instantaneous fluid velocity u in streamwise direction.*

Regime	$\lvert\mathbf{g}\rvert\, H\,/\,U_b^2$	$(\rho_p - \rho_f)\,/\,\rho_f$	Fr	Sh	Sh_{crit}
MA	1	1	1	0.039	0.034
MB	9.81	0.2	2	0.020	0.034

Table 3.3: *Dimensionless parameters characterizing the two regimes in the studies with multiple mobile particles.*

Regime MA, the Shields number is larger than the critical value of $Sh_{crit} = 0.035$ according to [141], while in Regime MB the value of Sh is slightly lower than the threshold for incipient sediment motion. Hence, sediment transport is expected for Regime MA and no sediment transport for Regime MB.

The final simulation matrix for the investigations of mobile particles and the various collision models is shown in Tab. 3.4. With this matrix all aspects of the collision modeling can be investigated. The idea is to increase the complexity of the collision modeling by first using a basic model and further on including effects like lubrication and tangential interaction. The parameters in the ACTM and in the ATFM are set to $e = 0.97$, $\Psi_{in}^{crit} = 0.95$ and $\mu = 0.15$, respectively. These are typical values for rough glass spheres as obtained in the experiment of Joseph & Hunt [79], for example.

Case	Regime	RPM (3.4)	ACTM (3.5a)	LM (3.8)	ATFM (3.7)
1	MA	x	-	-	-
2	MA	-	x	-	-
3	MA	-	x	x	-
4	MA	-	x	x	x
5	MB	x	-	-	-
6	MB	-	x	x	x

Table 3.4: *Overview of simulations with multiple mobile particles to investigate the impact of the collision model on near-bed particle transport. Numbers in brackets refer to the respective equations. Note that when all three sub-models are used, i.e. ACTM + LM + ACTM, the label ACM is employed below for conciseness.*

a)

b)

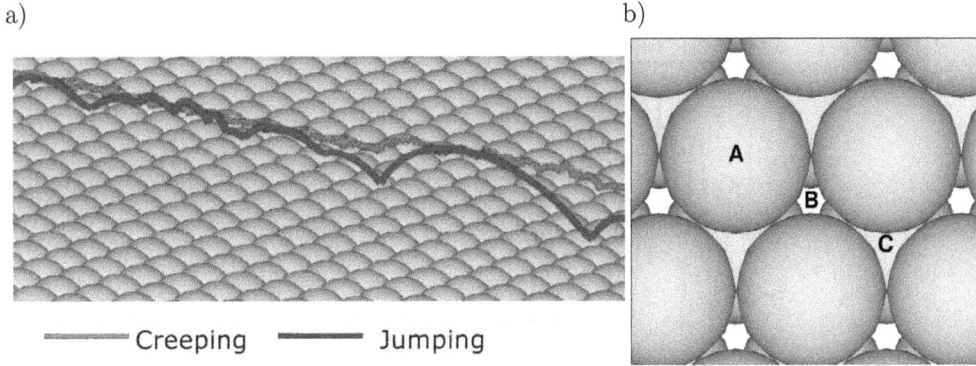

Creeping ——— Jumping

Figure 3.6: *a) Typical trajectories for creeping an jumping motion of a particle over the rough bed. b) Local extrema of elevation of a hexagonally packed bed. The distance between Points B and C is used as criterion to distinguish between creeping and jumping motion.*

3.4 Results for single mobile particles

3.4.1 Classification of the types of motion

Lajeunesse *et al.* [96] distinguish between three different kinematic states: resting, creeping, and jumping. A typical trajectory of a particle in creeping motion and in jumping motion computed in the present study is shown in Fig. 3.6a. In the following, critical parameters are introduced to distinguish between these three types of motion. First, the mean free path of a particle between two subsequent collisions is defined as the path integral of the linear particle velocity, i.e.

$$L_{trans}^{(i)} = \int_{t_c^{i-1}}^{t_c^i} |\mathbf{u}_p(t)| \, \mathrm{d}t \quad , \tag{3.14}$$

where t_c^{i-1} is the time of the previous collision and t_c^i the time of the current collision. To distinguish between resting and creeping motion, the average velocity between two subsequent collisions of the particle with the rough wall is taken as the critical parameter. It is defined by

$$u_{bal}^{(i)} = \frac{L_{trans}^{(i)}}{t_c^i - t_c^{i-1}} \quad , \tag{3.15}$$

where the index *bal* refers to "ballistic". Since a large increase in the jumping length was observed when the velocity exceeds $u_{bal}^{crit} = 0.5 \, u_\tau$, this value is taken here as a threshold between resting and creeping motion.

Creeping is characterized by a rolling or sliding motion of the particle and a strong interaction with the sediment bed, i.e. a large number of collisions per distance propagated. Jumping, in turn, involves only weak interaction with the sediment bed as most of the points on the trajectory are remote from the sediment. Hence, the threshold to distinguish between creeping and jumping is defined here in terms of the mean free path between two collisions, L_{trans}, which depends on the local geometry of the sediment bed. The critical value is chosen equal to the distance between two possible local minima of elevation experienced by a particle resting on the fixed bed. For the present hexagonal packing, these are the points

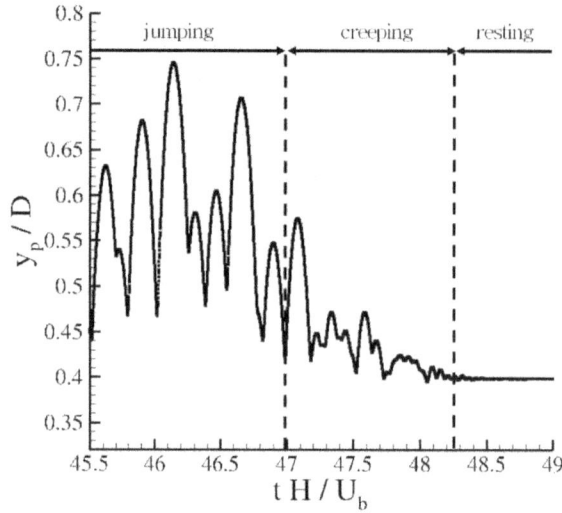

Figure 3.7: *Subdivision of a particle trajectory into lapses of time assigned to jumping, creeping and resting, respectively.*

B and C in Fig. 3.6b. Therefore, L_{trans}^{crit} is set equal to $D\sqrt{3}/3$. This procedure is similar to the method applied by Lajeunesse *et al.* [96] who used a value of $0.9\,D$ for a cubical packing. Since all three types of motion may occur during the motion of the particle over the fixed bed, the computed trajectory is subdivided into lapses of time attributed to the different states of motion, as shown, for example, in Fig. 3.7. Subsequently, selected physical parameters, such as the linear and angular velocity components, are conditionally averaged over the respective lapses of the individual particle trajectories, and over all trajectories, to determine the respective quantities during the different kinematic states.

3.4.2   Jumping state

To quantify the impact of the collision models (RPM, ACTM + LM and ACM) in the jumping state, Probability Density Functions (PDFs) of the angular velocity in spanwise direction, $\omega_{p,z}$, and the jumping height of the particle, $H_{bal} = \max\{y_p(t)/D, t_c^{i-1} < t < t_c^i\}$, are shown in Figs. 3.8a and 3.8b, respectively. In this regime, the PDF of $\omega_{p,z}$ is close to Gaussian for the three collision models employed here. This is not unexpected since the interaction with the sediment bed is comparatively small so that the PDF of $\omega_{p,z}$ is influenced mostly by the viscous stresses acting on the particle. With increasing height above the sediment bed the velocity gradient diminishes so that the viscous forces decrease resulting in a lower angular velocity of the particle. If a tangential model is employed (ACM), the angular velocities increase because in addition to the interaction with the fluid angular momentum is transferred during the collisions. Fig. 3.8b shows the PDF of the jumping height of the particle, H_{bal}, in the jumping regime. The trajectories obtained with the ACM on average have slightly smaller jumping heights than obtained for the other cases. The position of the peaks of the PDF shown in Fig. 3.8b agrees with the observation from Fig. 3.8a, as more frequent jumps of low height lead to an increase of angular velocity.

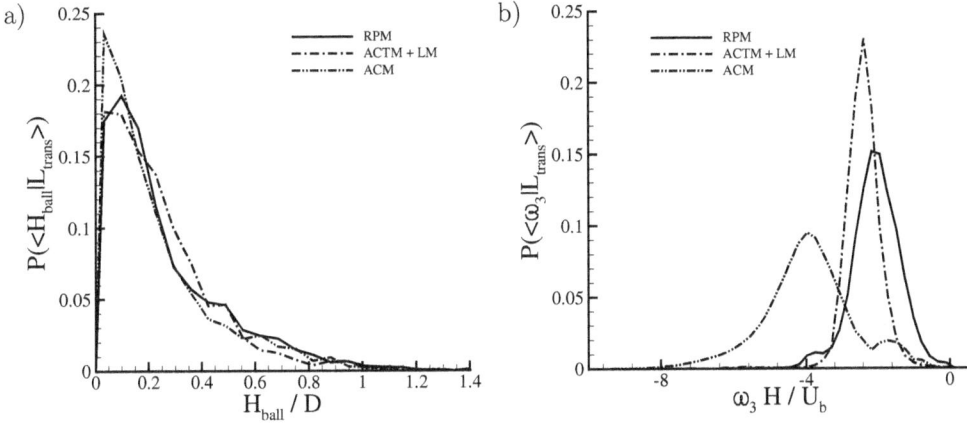

Figure 3.8: *PDFs of the jumping state. a) Angular velocity (rotation around spanwise axis of the particle. b) Jumping height of the particle.*

3.4.3   Creeping state

When the particle creeps over the fixed sediment bed, interaction with the immobile particles is frequent, so that the collision model is supposed to have an even larger influence. This indeed is visible when determining the PDFs of the streamwise velocity of the particle, u_p, the angular velocity, $\omega_{p,z}$, and the PDFs of the wall-normal position of the particles, which are shown in Fig. 3.9a, 3.9b and 3.10a, respectively. Fig. 3.9a shows that the mean linear velocity is substantially lower with the ACM compared to the RPM. This is an effect caused by the repulsive force range of $S = 2h$ used in this model. It makes the mobile particle hover at a distance of about $2\,h$ above the sediment bed revealed by the PDFs of the wall-normal position of the particles. Therefore, the particles are protruding deeper into the turbulent boundary layer than with the ACM, which in turn leads to higher pressure forces and viscous forces on creeping particles if the RPM is used.

A somewhat similar effect can be seen in Fig. 3.9b. Again, if the RPM is used, the particle has higher rotational velocity, so that the model yields larger kinetic energy, angular as well as translational. Moreover, the use of the tangential model (ACM) leads to an increase of rotational velocity. This increase is due to the fact that the spinning motion caused by the viscous stresses can actually be translated into a rolling motion by the tangential interaction with the fixed particles due to friction.

The experiments of Lajeunesse *et al.* [96] show that if a particle rolls along the sediment bed it exhibits a strong interaction with the bed and the jumping length approaches zero. The circumferential velocity is addressed by the length

$$L_{circ}^{(i)} = \int_{t_c^{i-1}}^{t_c^i} |R_p \left(\boldsymbol{\omega}_p \times \mathbf{n}_{pq} \right)| \, \mathrm{d}t \quad , \tag{3.16}$$

between two collisions at t_c^{i-1} and t_c^i respectively, where $\mathbf{n}_{pq}$ is the unit vector pointing from the center of mass of the mobile particle to the contact point with the fixed particle. When the particle is rolling, the value of $L_{circ}^{(i)}$ should be equal to the integral of the linear velocity

a)

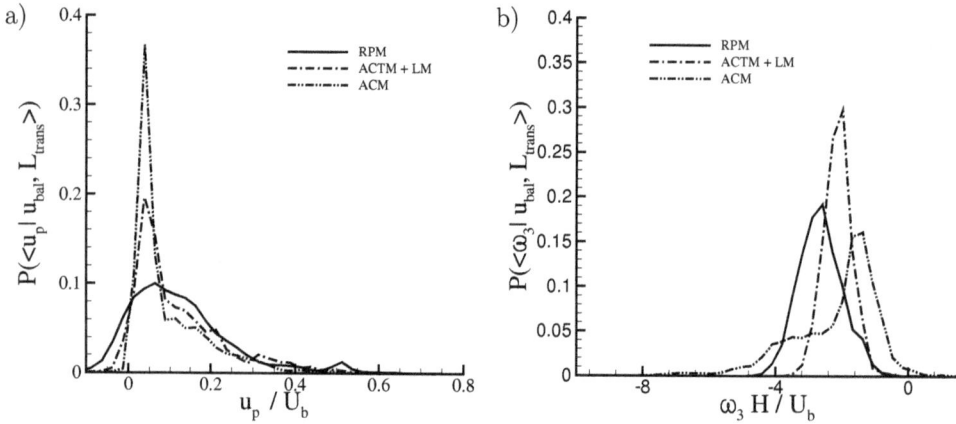

b)

Figure 3.9: *PDFs of a) the streamwise velocity of the particle and b) of the angular velocity of the particle (rotation around spanwise axis) for creeping motion.*

$L_{trans}^{(i)}$ according to Equation (3.14). Hence, the slip length

$$L_{slip}^{(i)} = L_{circ}^{(i)} - L_{trans}^{(i)} \qquad (3.17)$$

is suitable to quantify the deviation from pure rolling. Averaging over all time intervals with creeping motion yields the corresponding mean value L_{slip}. The result is displayed in Fig. 3.10b for different values of L_{trans}, i.e. the distance between two collisions. For physically realistic collision models also including tangential forces, L_{slip} should be close to zero. Obviously, with decreasing L_{trans} there is a considerable slip of the surfaces at the contact point if a model for the tangential forces is absent. It is important to notice that by construction the RPM leads to a constant rotational motion of the particle, independent of the collision process.

To investigate the impact of the effects mentioned above, the overall percentage of time that a particle spends in one of the three kinematic states was calculated for the three different density ratios ρ_p/ρ_f employed (Tab.3.2). Fig. 3.11 shows that with increasing particle density the impact of the collision model increases. Light particles spend most of the time in the jumping state, i.e. far from the lower boundary, with rare contact. Their trajectory hence is not as dependent on the collision model as the trajectory of heavy particles. For the latter, the time spent in creeping state or resting state increases drastically. As discussed in the context of Fig. 3.9a the particle exhibits a lower protrusion into the mean flow with the ACM because no *a priori* repulsive range S is employed. The rebound height therefore is lower because of the smaller hydrodynamic forces acting on the particle in this case. The tangential model, on the other hand, leads to longer periods of creeping before the particle comes to rest again. While the particle stays in a constantly spinning motion without tangential forces due to the vertical velocity gradient and the viscous forces from the flow, this is suppressed with the tangential force model. With such a model being present, the angular momentum exerted by the fluid on the particle generates a rolling motion.

Calculating the resulting Shields parameter Sh of the different density ratios gives values of 0.0171 for the light particles, 0.0057 for the intermediate density ratio, and 0.0034 for the heavy particles (Tab. 3.2). This is a parameter range that compares very well with the

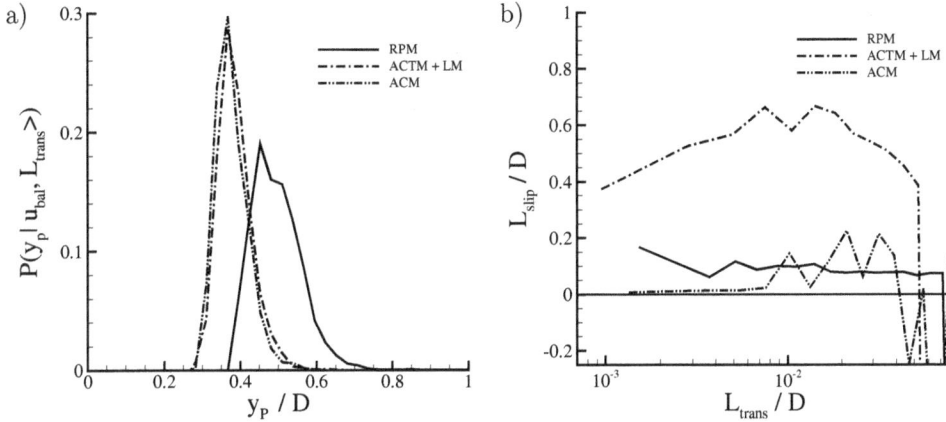

Figure 3.10: *Single particle in creeping motion. a) PDFs of the wall-normal position of the particles and b) the slipping length of the particles according to (3.17).*

experiments performed by Coleman [40] and Fenton & Abbott [55]. In these experiments, particles were placed on top of a hexagonal packing of monodisperse spheres, which gives an exposure of 82%. Afterwards, the fluid conditions were adjusted until incipient motion of the fully exposed particle occurred. For a particle Reynolds number of $D^+ = 23$, a critical Shields parameter $Sh_{crit} = 0.004$ was found. For the numerical study presented here, the heaviest particle is slightly below the experimental threshold. For the given physical parameters of this setup, the particle should therefore hardly be in motion. This behavior is represented substantially better by the ACM compared to the other models investigated.

3.5 Results for many mobile particles

3.5.1 Sampling strategy

In this section simulations with many mobile particles are reported for the configuration described in Section 3.3.3, a situation representing bed-load transport of sediment. Statistical data of fluid and disperse phase obtained from simulations with different collision models are presented. After the initialization phase, the flow was advanced in time until fluid turbulence and particle structures were developed. Afterwards, samples were taken until no significant change of the second-order statistics was visible any more when further increasing the sampling time. This procedure resulted in averaging over more than 300 time units H/U_b and corresponds to approximately 50 flow-trough times L_x / U_b. The mean velocities and the Reynolds stresses were obtained by point-wise temporal averaging and spatial averaging in homogeneous directions. This was done here without special treatment of Eulerian grid cells lying in the interior of solid particles.

3.5.2 Flow field statistics for light particles

The mean velocity and the Reynolds stresses for the simulations of Regime MA using different collision models are shown in Fig. 3.12. For reference, the simulation of the flow over the

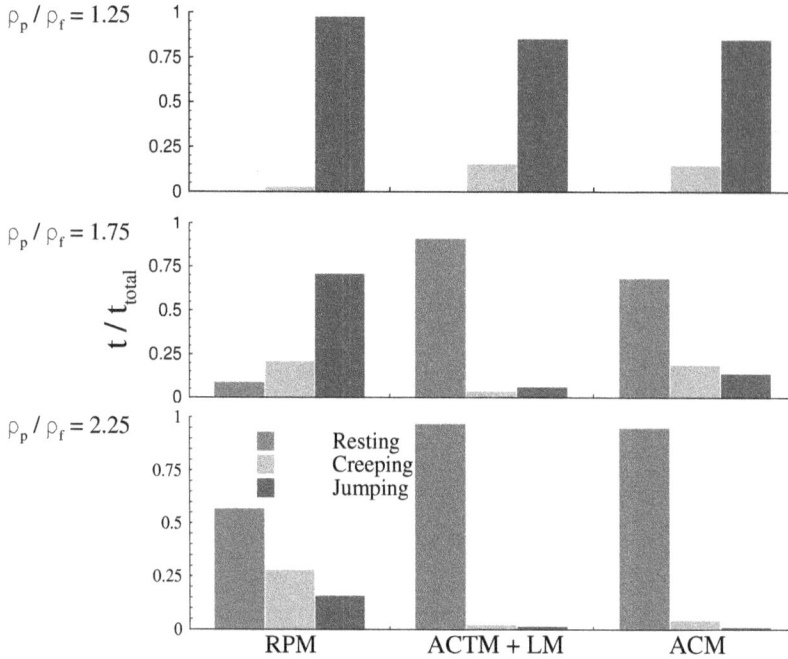

Figure 3.11: *Overall fraction of time the particle spends in different kinematic states (1. bar: resting, 2. bar: creeping, 3. bar: jumping, from left to right).*

rough bed without mobile particles is shown in the diagrams as well. Obviously, the profiles of mean velocity and Reynolds stresses for simulations with and without mobile particles are considerably different. A significant velocity defect in comparison to the single phase flow (label unladen) can be observed in the near-bed region for all simulations with moving particles. This is expected since the additional particles, mobile or not, constitute additional obstacles to the flow. The peak in $\langle u'u' \rangle$ is substantially broadened in all simulations with particles compared to the simulations with rough bottom wall alone. Employing the RPM the maximum of $\langle u'u' \rangle$ is slightly lowered in contrast to the simulation without particles. The result is similar to the ones obtained by Chan-Braun *et al.* [36], where the RPM was used for collision modeling throughout.

3.5.3   Particle statistics for light particles

This section focuses on statistical results for the disperse phase in the simulations with $Fr = 1$ (Regime MA). The averages of quantities related to the particles on top of the fixed bed are carried out for discrete bins in the wall normal direction. On the one hand, discrete binning of data tends to smooth out the profiles. On the other hand, if bins are too small a very large total number of samples is required to ensure statistically converged data. As a compromise to ensure a sufficient number of samples per bin and the representation of sharp gradients in the dispersed phase, the width of the bins was set to $D / 10$ here. A particle is assigned to a bin if its center point is located inside.

The PDFs of the wall normal position of the particles are shown in Fig. 3.13a. Due to the choice of the forcing range, $S = 2\,h$, the particle positions obtained with the RPM are

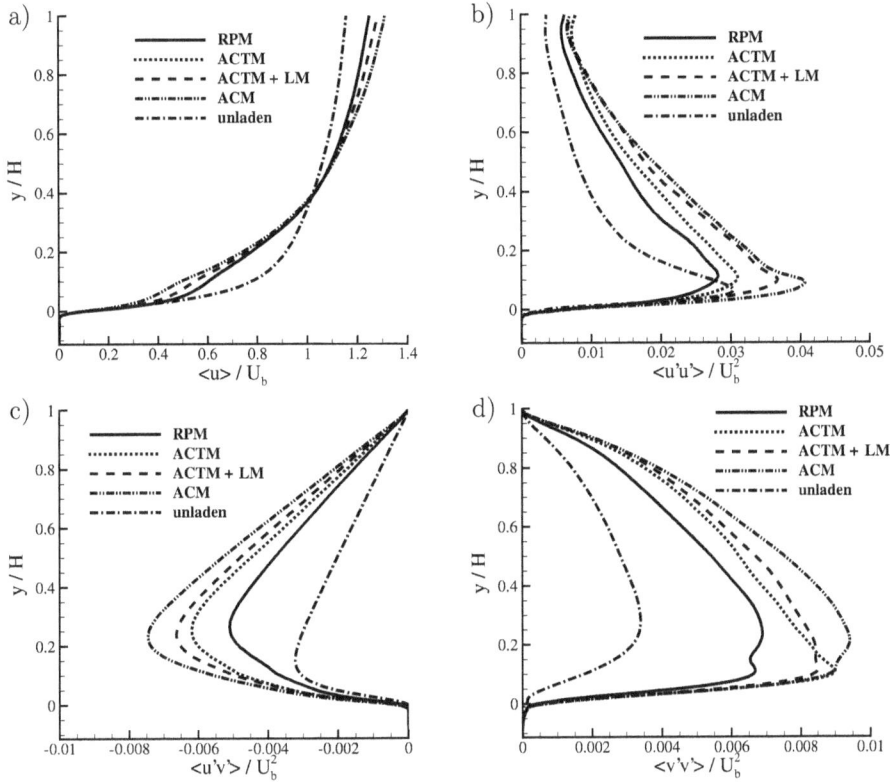

Figure 3.12: *Mean velocity and Reynolds stresses for the simulations with Fr = 1 and various collisions models. The results for a rough channel without mobile particles are shown for comparison (unladen).*

generally higher than with the ACTM directly affecting the particle protrusion which is an important characteristic influencing mobilization to a large extent [55]. Coupling the ACTM with the lubrication model (LM) does not substantially affect the wall normal distribution of the particles. When tangential forces are added, the particles are located at slightly lower positions, and the near wall peak increases by about 15 %.

The mean particle velocity in streamwise direction is shown in Fig. 3.13b. Due to the very small number of samples for $y_p / D > 3$, indicated by Fig. 3.13a, the curves are not displayed for this region. While for $y_p / D > 1.5$ the profiles are nearly identical, significant differences of the particle velocity exist in the region $y_p / D < 1.5$. Again, the profile obtained with the RPM has an offset due to the repulsive range. The velocity recorded in the near-wall region is higher due to the larger exposure of the particles to the flow and the lack of tangential forces, caused by the fact that the particles tend to hover above the sediment, as noticed in Section 3.4 above. When instead of the RPM the ACTM is used without LM and ATFM, the particle velocity is smaller due to the larger interaction with the immobile bed, because the particles get closer to it in that case. Additionally, the interaction of particles above those touching the sediment bed with the ones below is smaller, so that the upper particles remain faster. When adding the lubrication model the particle velocity is decreased substantially. The LM only acts in normal direction but normal collisions with a rough bed can have a streamwise component. Another perspective is to observe that the LM is dissipative, and

a)

b)

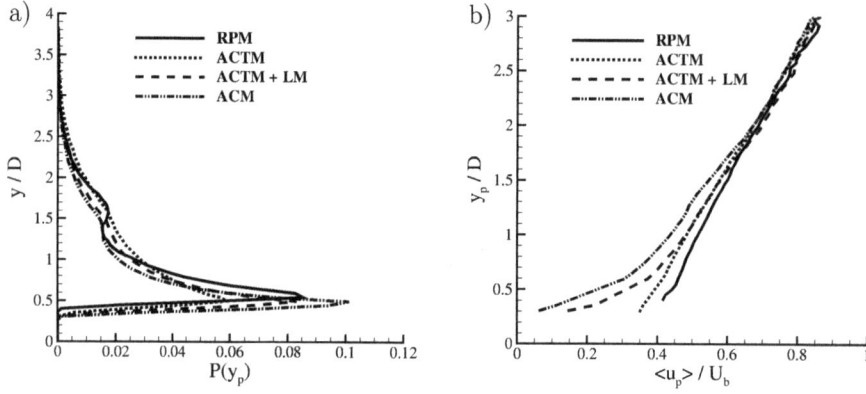

Figure 3.13: *Impact of the collision model for the simulations with $Fr = 1$. a) Probability density function of the wall normal position of the particle centres and b) mean particle velocity in streamwise direction for different heights.*

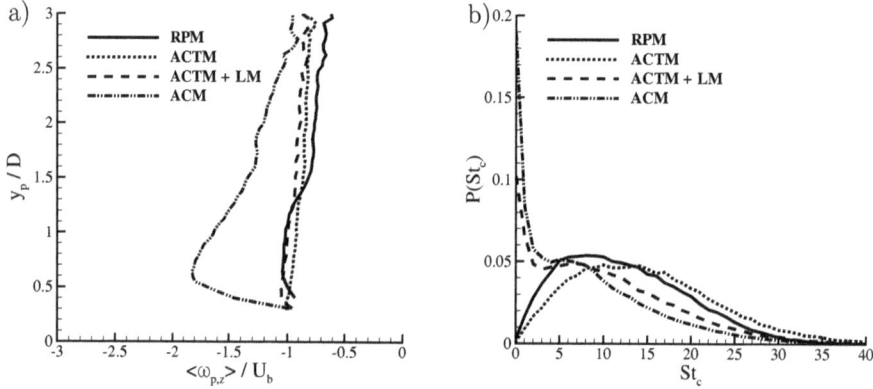

a)

b)

Figure 3.14: *Simulations with $Fr = 1$ and various collision models. a) Mean angular velocity of the particles (rotation around spanwise axis) for different heights and b) probability density function of the Stokes number St_c based on the particle velocity right before the impact. Both, particle-particle and particle-wall collisions are accounted for.*

although it acts only over a very small distance it drains energy from the particles. Still larger is the modification introduced by the entire ACM, driving the average particle velocity almost to zero at the bottom. Even at a distance of D over the lowest possible position which can be reached by a mobile particle the difference in particle velocity is 20 % compared to the RPM.

The impact of the tangential force model can be further appreciated in Fig. 3.14a showing the mean angular velocity in spanwise direction of the particles, $\langle \omega_{p,z} \rangle$. It can be seen that the application of the ATFM yields a significant rotation of the particles with a strong peak of $\langle \omega_{p,z} \rangle$ close to the bottom. Also particles farther away from the bed, at higher elevation, rotate substantially faster. With normal forces only, i.e. ACTM and ACTM + LM, the rotation rate is fairly uniform in y and mainly induced by the shear of the carrier fluid. Using the RPM results in similar rotation close to the bed with a distinct kink about one diameter higher. This again is due to the fact that the particles close to the bed hover above

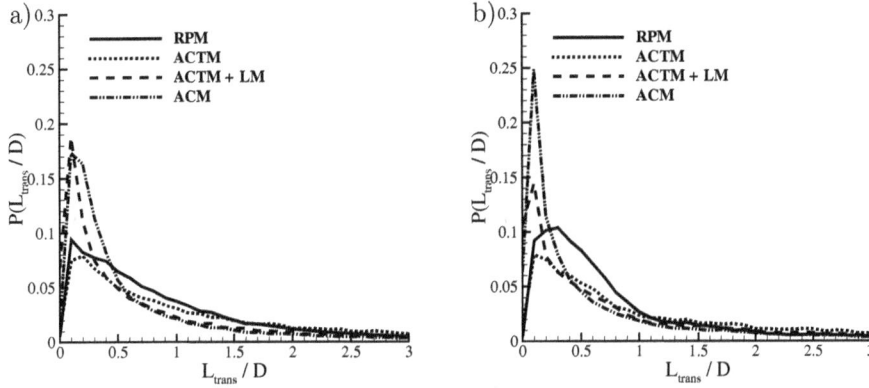

Figure 3.15: *PDF of the mean free path L_{trans} for the simulation of light particles ($Fr = 1$). a) Particle-particle collisions only and b) particle wall collisions only.*

the fixed particles and rotate due to the local gradient of the mean velocity. Furthermore, the particles move differently in terms of their collective behavior, hinted at by the particle positions in Fig. 3.13a above and discussed below in more detail.

To measure the intensity of the particle contacts when using the different collision models, the Stokes number

$$St_c = \frac{1}{9}\frac{\rho_p}{\rho_f}\frac{D\,u_{in,c}}{\nu_f} \quad . \tag{3.18}$$

is computed. In general, the Stokes number is defined as the ratio of the hydrodynamic response time of the particle $\tau_p = \rho_p\,D^2\,/\,(18\,\rho_f\,\nu_f)$ to a characteristic flow time τ_f. It is based here on the relative velocity at the beginning of the "dry" collision, i.e. at the beginning of the direct contact of the particles, so that τ_f is set to $D\,/\,(2\,u_{p,in})$. This is in contrast to the definition in [87] where St is based on the velocity of the particles right before the deceleration due to the viscous forces in the gap between the surfaces. This definition, however, can only be applied to binary collisions in quiescent fluid and not to the present case with multiple simultaneous collisions which motivates the use of St_c from (3.18) here. The PDF of St_c, shown in Fig. 3.14b, was computed for all particle-particle and particle-wall collisions with the width of the bins being equal to 1. For the RPM and the ACTM, both only accounting for normal contact forces and disregarding lubrication, the collision intensity is fairly similar. For the cases with lubrication forces included, ACTM + LM and ACM, collisions with lower Stokes numbers are substantially more frequent. Lubrication forces are always dissipative and decelerate the particle for both, approach and rebound, thus reducing their relative velocity. This is most pronounced for the complete model, ACM. Compared to RPM, the probability of $St_c \approx 15$ is reduced to less than 50 % and the peak around $St_c \approx 0$ increases substantially, while just not present with the RPM. The mean free path L_{trans} of the particles between two subsequent collisions was computed according to (3.14). In Fig. 3.15a, this quantity is displayed for particle-particle collisions and in Fig. 3.15b for particle-wall collisions. In only a few cases the free path is larger than one particle diameter. Similar to the mechanisms discussed in Fig. 3.14b, L_{trans} is significantly reduced for the cases where a lubrication model for the unresolved viscous forces is employed.

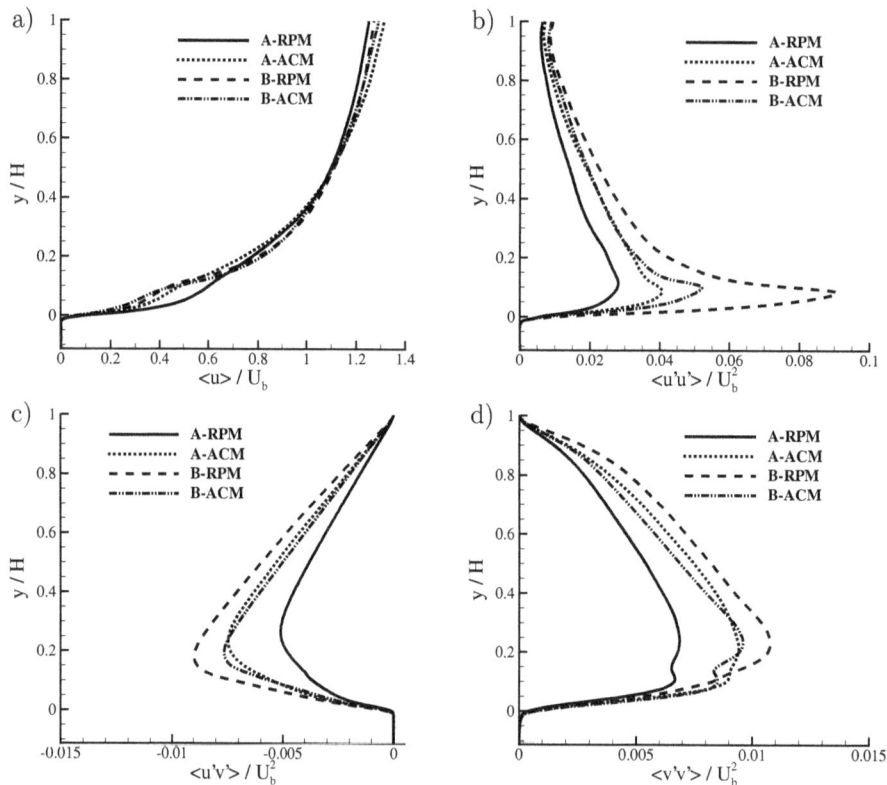

Figure 3.16: *Impact of the collision model comparing simulations with $Fr = 1$ and $Fr = 2$. Mean velocity and Reynolds stresses for various collisions models.*

3.5.4 Impact of the collision model for heavy particles compared to light particles

In this section the impact of the type of collision model applied is investigated for heavier particles by comparing simulations with $Fr = 2$ (Regime MB in Tab. 3.3) with the previous results for $Fr = 1$. To focus the discussion, only the cases RPM and ACM are retained since these lead to the largest differences in the previous section. The statistics of the fluid phase shown in Fig. 3.16 indicate that the flow is very sensitive to the type of collision model for both Regimes MA, and MB. For example, in the case of full modeling with the ACM, turbulent fluctuations $\langle u'u' \rangle$ are increased by approximately 40 % in comparison with the simulations using the RPM. The differences resulting from the collision model are somewhat smaller for the other Reynolds stresses in Regime MB. When comparing these data it is important to note that the collective behavior of the particles can be, and indeed is, different in the cases shown below.

The PDF of the wall-normal position of the particles in the two regimes as well as the mean particle velocity in streamwise direction are shown in Figs. 3.17a and 3.17b, respectively. Since the Froude number is relatively large in Regime MB, the particles stay on average very close to the sediment bed. Only few particles reach heights of more than one particle diameter. As mentioned above, the particle positions obtained with the RPM are generally higher than with the ACM due to the choice of the force range $S = 2\,h$ in the RPM

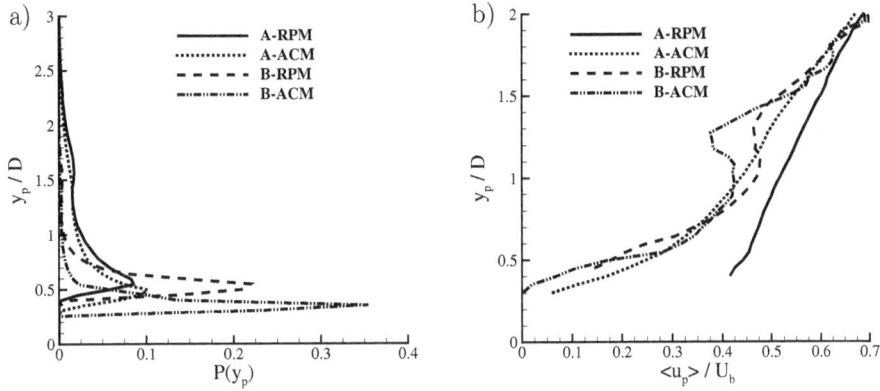

Figure 3.17: *Comparison between simulations with $Fr = 1$ and $Fr = 2$. a) Probability density function of the wall normal position of the particle centers and b) mean particle velocity in streamwise direction over height.*

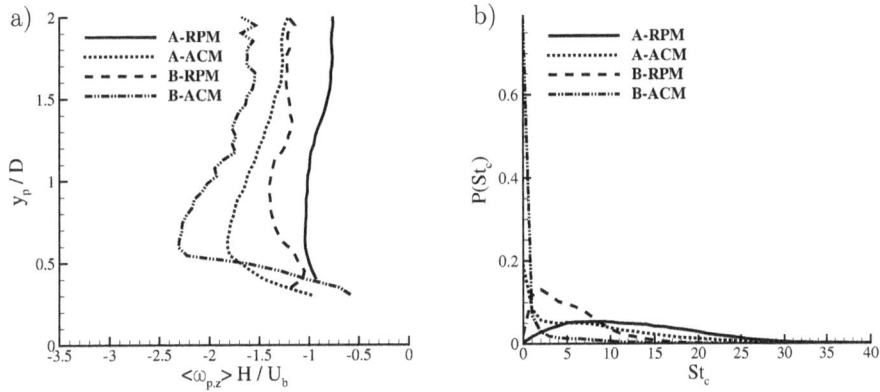

Figure 3.18: *Comparison between simulations with $Fr = 1$ and $Fr = 2$. a) Mean angular velocity of the particles (rotation around spanwise axis) for different heights and b) probability density function of the Stokes number St_c based on the particle velocity right before the impact, summing over both, particle-particle and particle-wall collisions.*

introducing a substantial difference here. The particle velocity near the bed in Fig. 3.17a exhibits a less drastic change when switching from RPM to ACM in the case with $Fr = 2$ compared to the case with $Fr = 1$. Nevertheless, the difference is still noticeable, in particular for $1 < y_p / D < 1.5$.

The mean angular velocity around the spanwise axis and the Stokes number based on the impact velocity of the particles are displayed in Fig. 3.18. With the ACM, the angular velocity of the particles near the bed is higher than with the RPM. Furthermore, the shape of the curve is qualitatively different between the two models. The average Stokes number at the particle impact is reduced for Regime MB in contrast to the simulations in Regime MA. Since the particles are relatively heavy they are not very often lifted off by the fluid from the fixed layer. Hence, their relative velocity right before collisions is small. Again, a substantial difference in the result is to be observed when changing from RPM to ACM.

3.5.5   Average particle velocity and Shields number

In this section the impact of the collision model is underlined by comparison of the numerical data with experimental data and observations. The volume-averaged mean particle velocity, $\langle u_p \rangle_{x,y,z,t}$, is proportional to the mass transport by the dispersed phase and is reported in Tab. 3.5. The critical Shields number for the present setup is $Sh_{crit} = 0.035$. Therefore, no significant sediment transport should occur with $Sh = 0.02$ in Regime MB (Tab. 3.3), so that a small volume-averaged particle velocity indicates physically plausible results. For Case 5 (Regime MB and RPM) the normalized particle velocity is $\langle u_p \rangle / U_b = 0.245$. This value is reduced to 0.083 for Case 6 (Regime MB and ACM), where all physical effects are accounted for in the collision model. Hence, the more complex model yields higher physical realism as the mean particle velocity is smaller. Analogously, the average particle velocity is reduced from 0.551 for Case 1 (Regime MA and RPM) to 0.353 for Case 4 (Regime MA and ACM) so that a significant impact of the collision model on sediment transport is also found for the simulations with lighter particles.

Case	1	2	3	4	5	6
$\langle u_p \rangle_{x,y,z,t} / U_b$	0.511	0.483	0.438	0.353	0.245	0.083

Table 3.5: *Computed particle velocity in the investigations with multiple mobile particles. The velocity is averaged over the entire domain and in time. The cases correspond to the definitions in Tab. 3.4.*

3.5.6   Fluid–worked structures of the sediment

In this final section, the instantaneous flow patterns of the dispersed phase are investigated. In Fig. 3.19, a typical snapshot of the particle distribution is shown for the simulations of lighter particles with the RPM and the ACTM. The particles are colored by their instantaneous velocity in streamwise direction. Fig. 3.19a reveals the presence of dune-like structures perpendicular to the flow in this case. Nearly all particles are colored in red, indicating that this dune is sliding over the fixed bed. This is in contrast to moving dunes investigated in experiments, where only the particles along the upwind front of the dune are gradually mobilized [9]. The unphysical behavior observed with the RPM is also present in results for Regime MB displayed in Fig. 3.20a. Again, a dune is formed in spanwise direction and slides over the bed. In contrast to the simulation in Regime MA (Case 1), the velocity of the particles is significantly reduced but not zero, so that they still hover above the fixed particles without direct surfaces contact.

When employing the ACM, longitudinal ridges of particles are created in Regime MA as well as in Regime MB (Fig. 3.19b and 3.20b). These are related to a secondary flow composed of pairs of counter-rotating vortices with their axis in streamwise direction. Such a pattern is often observed in natural river beds and experiments with low mobility of the particles and low mass loading [9, 110, 120, 142] as in the present configuration. They are periodic in spanwise direction with a typical period length of about $\lambda_z \approx 2\,H$ [110]. In the present case the distance between the ridges is $1.5\,H$ caused by the size of the computational domain in spanwise direction being $L_z = 3\,H$, which enforces the evolution of two counter rotating vortices in streamwise direction with a wavelength of $\lambda_z = 1.5\,H$. This can be improved by a larger domain but would not add much here. The goal of the

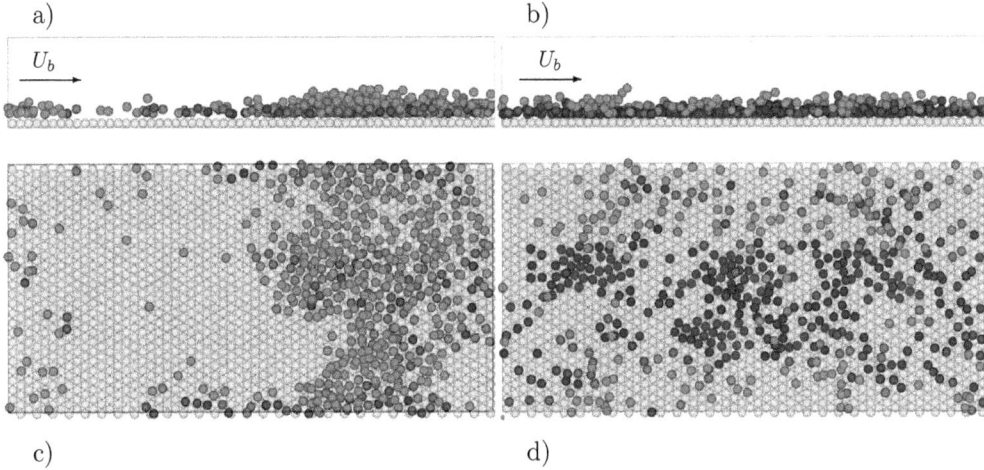

Figure 3.19: *Snapshot of the particle structures in a turbulent open channel obtained with with light mobile particles (Fr = 1, Regime MA). Left (a, c) Case 1 using the RPM, right (b, d) Case 4 using the ACM. The upper pictures a) and b) show side views at the same instance of the respective top views in c) and d). The Particles are colored by their velocity in streamwise direction. Blue color: $u_p = 0$, red color: $u_p = 0.5$ (Figure taken by permission from [86]).*

present study is met by demonstrating a substantial impact of the collision model on the three-dimensional simulation of moving sediment particles. Further quantitative analysis of the physical mechanisms of developed sediment transport is beyond the scope of this chapter but it is addressed in Chapter 4 and 5, where results are presented obtained with significantly enlarged computational domains. The results of the present chapter stand in pronounced contrast to the two-dimensional simulations performed by Papista *et al.* [125]. In this study the authors conclude that the collective behavior of the particles remains qualitatively the same if a collision model with or without tangential forces is used, the present study clearly demonstrates that this does not hold for general three-dimensional simulations.

3.6   Concluding remarks

For sediment transport in bed-load mode the collisions between particles, either both moving or one fixed one moving are a key issue. Hence, their adequate modeling is of predominant interest for the understanding of this type of flow, as well as for its successful simulation. The present study demonstrates this fact by employing different collision models in a systematic way. They range from a very simple model just constituted by a repulsive force (RPM) to the newly developed Adaptive Collision Model (ACM) of Kempe & Fröhlich [87] accounting for lubrication, proper stiffness and damping during the phase of direct surface contact, as well as realistic tangential forces. The ACM so far has only been applied for single collisions in laminar flow and its application is extended here to multiple simultaneous collisions and to the contact-dominated regime.

In the present chapter, the ACM was employed for two situations, a single particle traveling over a fixed sediment bed, and a situation with bed load transport of many mobile particles. It was observed that the collision model can be applied in a straight forward manner to the complex situation of bed load transport without any further modification. The re-

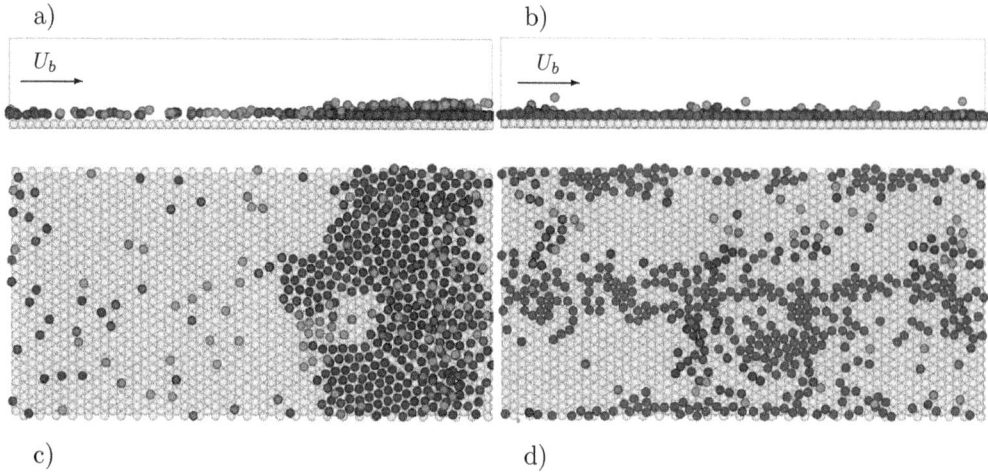

Figure 3.20: *Snapshot of the particle structures in a turbulent open channel obtained with with light mobile particles (Fr = 2, Regime MB). Left (a, c) Case 1 using the RPM, right (b, d) Case 4 using the ACM. The upper pictures a) and b) show side views at the same instance of the respective top views in c) and d). The Particles are colored by their velocity in streamwise direction. Blue color: $u_p = 0$, red color: $u_p = 0.5$ (Figure taken by permission from [86]).*

sults obtained with the full ACM are physically more realistic than with simpler models. This was demonstrated by means of statistical data for fluid as well as for particle motion showing significant dependency on the collision model. Detailed discussion and comparison to experimental observations demonstrate that the new ACM is able to represent bed-load transport with considerably improved realism compared to other models. This improvement is achieved without noticeable increase of computational cost. Furthermore, physically relevant information on the situation of a single particle traveling over a rough bed was obtained. For bed-load transport of clouds of particles, the same computational domain was selected for reasons of comparison and substantial dependency of the results on the collision model was observed as well. No experiments are available for this setting. The quantitative discussion of pattern formation in this type of flow is beyond the scope of the present chapter and requires substantially larger domains. This will be presented in the following two chapters. The sizable qualitative and quantitative differences observed in the various results reported in this paper demonstrate that the collision model has a critical impact in three-dimensional simulations of bed-load transport with spatially resolved particles. The recently proposed ACM shows superior performance and provides a realistic, cost-effective model for this situation.

4 Momentum balance in flows over mobile granular beds: a double–averaging analysis

4.1 Introduction

The approach of Direct Numerical Simulations (DNS) with spatially resolved solid-liquid interfaces using the Immersed Boundary Method (IBM) as described in Chaps. 2 and 3 provide a promising approach, to generate high resolution data in relation to both space and time with high fidelity. As outlined in Sec. 1.4, this chapter aims to convolute the enormous amount of data generated by DNS into a handy set of parameters applicable for engineering applications, up-scaling, and physical interpretations. For smooth-bed flows researchers and engineers may successfully apply the Reynolds-Averaged Navier-Stokes (RANS) framework, which deals with time- (ensemble) averaged variables, but involves no spatial averaging. Although these equations may still be used, in principle, for fixed rough-bed flows, their practical application is problematic due to complex boundary conditions. This inconvenience is removed by spatial averaging of RANS equations that produces double-averaged (in time and space) hydrodynamic equations [131, 67, 57]. This framework is known as the Double-Averaging Methodology (DAM). For a more general case of mobile rough-bed flows the conventional Reynolds-averaged hydrodynamic equations are not even applicable and thus need to be replaced with a more general form of the double-averaged hydrodynamic equations accounting for the effects of mobile boundaries. A detailed discussion of these issues is provided in [115, 116].

In this chapter, the DAM is applied to DNS data produced with the IBM. The dimensions of the computational domain resemble open-channel flows with small relative submergence. The domain bottom was a flat plane covered with one layer of hexagonally packed, monodisperse spheres fixed to the bed. The fixed particles were covered by 2000 mobile particles of the same size that were free to move. Two simulation scenarios at distinctly different values of the Shields parameter, one well above and another well below the nominal threshold of incipient motion, have been studied and are reported in this paper. The data analysis focuses on time-averaged and spatially-averaged flow quantities, including detailed assessment of the key terms of the double-averaged (in time and space) momentum balance equation formulated for mobile-bed conditions. As a first step, the effects of the averaging time and the averaging domain size as well as its shape were identified. It is found that for both simulation scenarios the flow is significantly influenced by moving bed particles, and this takes place in different ways depending on the Shields number. At low Shields parameter, longitudinal ridges of settled particles are formed introducing spanwise heterogeneities in

a)

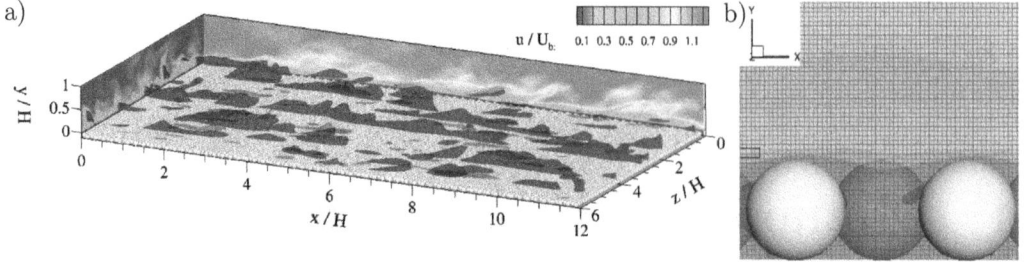

b)

Figure 4.1: *Illustration of the simulation setup (snapshots are shown). a) Computational domain of the unladen flow showing a contour plot of u/U_b on the sides of the domain, 3d-iso-surfaces of fluid fluctuations with $u'/U_b = -0.3$ in blue and $u'/U_b = 0.3$ in red inside the domain. b) Zoom of the fixed bed showing the background Eulerian grid and streamwise fluid velocity in a plane through the centers of the two front particles.*

Re_b	D^+	Re_τ	H/D	$N_{p,fix}$	$N_{p,mob}$
2700	19.2	177	9	6696	2000

Table 4.1: *Common physical parameters of the simulation scenarios.*

double-averaged flow quantities, e.g., flow areas of high form-induced (dispersive) stresses. At high Shields parameter, the distribution of mobile particles is close to homogeneous, with a layer of enhanced turbulence in the near-bed region. The DAM provides a reasonable reduction of the massive data sets, produced by the fully resolved simulations, to a manageable number of double-averaged parameters still yielding a detailed description of multi-phase flows including characterization of flow resistance and secondary currents.

Most of the results presented in the following sections are integrated in a manuscript submitted to the Journal of Fluid Mechanics [159].

4.2   Simulation scenarios and key parameters

4.2.1   Computational setup

A turbulent open-channel flow is considered in a rectangular computational domain with periodic boundary conditions in streamwise and spanwise directions, a free-slip condition at the top, and a no-slip condition at the bottom and the particle surfaces (Figure 4.1a). The sediment bed is constituted of a single layer of fixed spheres of diameter D, in hexagonal packing. The origin of the wall-normal coordinate y is set to the crest of the fixed spheres. The computational domain is $\Omega = [0; 12H] \times [-D; H] \times [0; 6H]$ with extents L_x, L_y, and L_z in streamwise, wall-normal, and spanwise direction, respectively. The submergence is $H/D = 9$. The bulk Reynolds number $Re_b = U_b H/\nu_f$ in the simulations is 2700, where U_b is the bulk velocity of the flow. The friction velocity is defined as $u_\tau = \sqrt{\tau_w/\rho_f}$, where the wall shear stress τ_w is obtained by extrapolating the total shear stress of an unladen flow down to $y = 0$. This shear velocity is used here to provide estimates of the roughness-regime, the grid resolution, and an *a priori* estimation of the Shields parameter for the particle-laden case as defined below in Secs. 4.2.2 and 4.2.3. The particle Reynolds number $D^+ = u_\tau D/\nu_f = 19.2$

L_x	L_y	L_z	Δ_x^+	D/Δ_x	N_l
$12H$	$H+D$	$6H$	0.86	22.2	1552

Table 4.2: *Common numerical parameters of the simulation scenarios.*

Scenario	ρ'	Sh/Sh_{crit}	$t_{init}\,[H/U_b]$	$t_{aver}\,[H/U_b]$
HP	1.15	0.7	40	175
LP	1.05	2.2	20	134

Table 4.3: *Specific parameters of the simulation scenarios.*

of the simulation indicates that it falls into the transitionally rough hydraulic regime [77]. The present setup is based on the flume design of [29], who used glass spheres of diameter $D = 11.1mm$ and oil as a fluid to obtain similar Reynolds numbers under experimental conditions. For convenience, common physical parameters are assembled in Table 4.1.

To resolve the viscous length scale at the particle surface, an equidistant Cartesian grid of $N_x \times N_y \times N_z = 2400 \times 223 \times 1206$ with constant isotropic cell size over the whole domain is employed, yielding total amount of 645 million grid cells and results in 22.2 grid points per particle diameter (Figure 4.1b). The resolution in terms of wall units is $\Delta_x^+ = u_\tau \Delta_x / \nu_f = 0.86$. A total number $N_l = 1552$ of Lagrangian marker points on the surface of a single particle was used. The most important numerical parameters are collected in Table 4.2. Except if stated differently, all variables are normalized with H and U_b.

To simulate a flow within and above a mobile granular bed, 2000 particles of the same diameter D as those of the fixed sediment bed were released in the interior flow. Initially, they sediment towards the bottom and are then transported by the fluid in bed-load mode. The Shields parameter [141] is the ratio of the shear stress eroding the sediment to the gravitational force which tends to stabilize the sediment and defined as

$$Sh = \frac{u_\tau^2 \rho_f}{(\rho_p - \rho_f)gD} = \frac{u_\tau^2}{\rho' gD} \quad . \tag{4.1}$$

Two different scenarios were studied: i) particles with a large relative submerged density $\rho' = (\rho_p - \rho_f)/\rho_f$ just below the threshold of motion Sh_{crit} on the one hand (case HP), and ii) particles with a lower density yielding $Sh > Sh_{crit}$ on the other hand (case LP), with the respective data shown in Table 4.3. Here, $Sh_{crit} = 3.4\,10^{-2}$ is the critical value of Shields parameter for incipient motion extracted from the diagram presented in [141]. Thus, according to the Shields parameter, case LP should correspond to a mobile bed while case HP should correspond to near-critical bed conditions. For this *a priori* estimation, u_τ as defined above was used. Furthermore, since the amount of mobile particles is equal to 30% of a full particle layer, particle protrusion can locally become relatively large, substantially enhancing the actual particle mobility [55], a situation similar to paved gravel bed streams as described in [126]. Thus, the Shields parameter does not account for protrusion effects and it serves as an indicative parameter for the present scenarios rather than an exact criterion of incipient motion.

The simulations were first run until the erosion and deposition rates were in equilibrium, which was verified using the method presented in [154]. The corresponding time is denoted t_{init}. Subsequently, the data for the statistical analysis were gathered over a duration termed

a)

b)

c)

d)

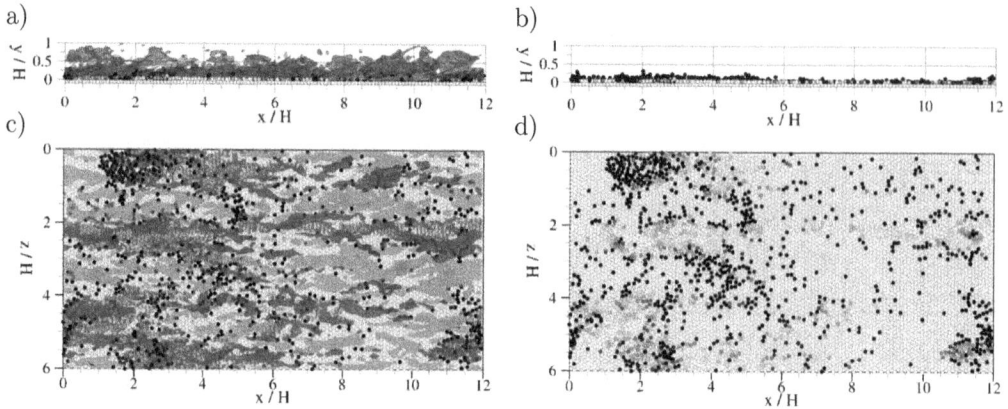

Figure 4.2:   *A snapshot of scenario HP. a) Side view of coherent structures visualized by u'/U_b (colouring as in Figure 4.1). b) Same as a) but without fluid information. c) Same as a) from the top. d) Same as in b) from the top. Particles in yellow: fixed, white: $|u_p| < u_\tau$, black: $|u_p| \geq u_\tau$.*

T_{sample}, with the corresponding values provided in Table 4.3 as a multiple of the bulk time unit $T_b = H/U_b$. The time interval between the samples collected for averaging was T_b. Preliminary flow analysis of the two scenarios considered in terms of one-point statistics were presented in [157] and [151]. In the remainder of this section, a description of the key flow features is given, before the DAM approach is applied to the data in the subsequent sections.

The simulation results were obtained on 256 cores of a SGI Altix at the Center for Information Services and High Performance Computing (ZIH), Dresden, and consumed in total more than 600,000 CPU hours, with additional information gained from simulations at JSC, Jülich.

4.2.2   Scenario with heavy particles

The heavy particles in scenario *HP* have a strong tendency to form streamwise clusters of resting particles as visible in Figure 4.2. This observation is in agreement with experiments of low transport rates, where ridges with an average spanwise spacing of $2H$ are reported [9, 84, 110, 142, 160]. In these references, it is argued that in between the ridges two counter-rotating cells of secondary flow develop, sweeping moving particles out of the troughs towards the ridges. The ridges of the present scenario show a similar pattern as observed in the experiments (Figure 4.2). Three characteristic large-scale clusters of particles with an average spacing of $2H$ can be observed in the snapshot, albeit with different intensities. While the structure at $z \approx 2H$ occupies the entire streamwise extent of the channel, a limited-size large cluster of resting particles with moving particles in the upstream front emerges at $z \approx 0H$. At $z \approx 4H$ the pattern is rather irregular, with some scattered particle clusters. The observed particle clusters clearly introduce spatial heterogeneity of the sediment bed in both, the streamwise and the spanwise direction. The cluster propagation speed differs by orders of magnitude from the fluid velocities yielding a significant separation between the morphological time scale and the turbulent time scale. Figure 4.2 illustrates that in response to the particle clusters by means of velocity fluctuations. Spatial scales of coherent fluid structures span from the particle scale to the full streamwise and wall-normal extent.

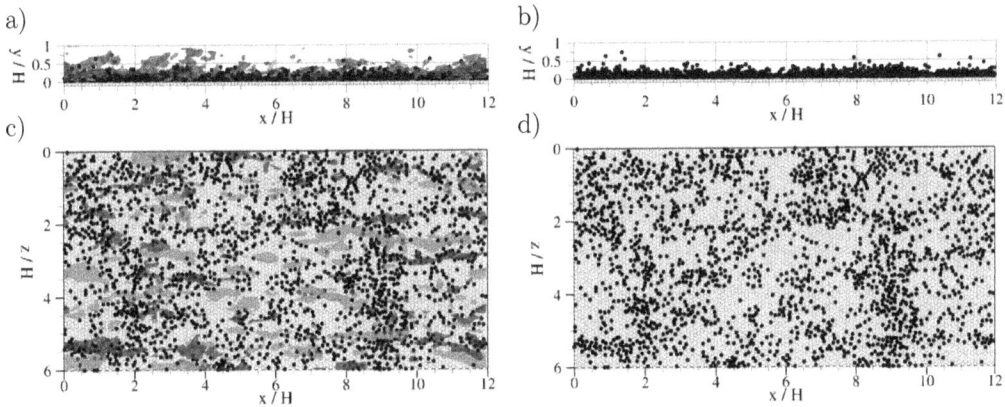

Figure 4.3: *A snapshot of scenario* LP. *a) side view of coherent structures visualized by* u'/U_b *(coloring as in Figure 4.1). b) Same as a) but without fluid information. c) Same as a) from the top. d) Same as in b) from the top. Particles in yellow: fixed, white:* $|\boldsymbol{u}_p| < u_\tau$, *black:* $|\boldsymbol{u}_p| \geq u_\tau$.

4.2.3   Scenario with light particles

Due to their much higher mobility, the light particles in scenario LP travel with a significant velocity over the fixed bed in streamwise direction as displayed in Figure 4.3 by their coloring. While doing so, they undergo frequent collisions with the bed and other mobile particles. It is also visible that the mobile particles form small-scale clusters, which are not stable in time. This behavior results in a fairly even distribution of the kinetic energy among the particles and no clusters of resting particles are observed. Hence, the particle distribution apparently is close to random and a separation of morphological and turbulent time scales does not occur. Despite the randomness of the distribution of the particles, increased large-scale coherent fluid structures are observed compared to the situation without mobile particles, although these are not as pronounced as in the scenario HP.

4.3   Double Averaging Methodology and its implementation

4.3.1   Definition of averaging operators

To interpret the present DNS data of turbulent bed-load transport, the double-averaged momentum equation for mobile bed conditions proposed by [112] is employed in the following. Full details of the derivation as well as considerations of alternative averaging procedures are given in this reference. Here, a short summary of the concept is provided in order to keep the chapter self-contained and to highlight the particularities introduced by the mobility of the sediment. Furthermore, this section and later parts of the chapter are conceived so as to highlight the bridge between the theoretical papers on DAM and the actual procedure of analyzing a given set of data.

The double-averaged momentum balance employs averaging operators similar to those used in studies of porous-media flows and multiphase flows [70]. Starting point is the distinction between the two phases, solid and liquid, by a clipping function γ equal to 1 in the fluid and

0 otherwise. Since particles are mobile, γ depends on time and space, i.e. $\gamma = \gamma(\mathbf{x}, t)$ with $\mathbf{x}$ the position vector. Temporal averaging of γ is indicated by an overbar and defined as

$$\overline{\gamma}^s(\mathbf{x}, t) = \frac{1}{T_0} \int_{\mathcal{T}(t)} \gamma(\mathbf{x}, t^*) \, dt^* \tag{4.2}$$

with

$$\mathcal{T}(t) = [t - T_0/2; t + T_0/2] \tag{4.3}$$

being the averaging interval of duration T_0[1]. The superscript s with the overbar indicates the so-called superficial average, to be distinguished from the so-called intrinsic average defined below. Here, the duration of the averaging, T_0, is set constant and selected as a parameter of the procedure. A similar averaging procedure is then defined for superficial averaging of γ with

$$\langle \gamma \rangle^s(\mathbf{x}, t) = \frac{1}{V_0} \int_{\mathcal{V}(\mathbf{x})} \gamma(\mathbf{x}^*, t) \, dV^* \qquad . \tag{4.4}$$

where

$$\begin{aligned} \mathcal{V}(\mathbf{x}) = &[x - L_{0,x}/2; x + L_{0,x}/2] \\ &\times [y - L_{0,y}/2; y + L_{0,y}/2] \\ &\times [z - L_{0,z}/2; z + L_{0,z}/2] \end{aligned} \tag{4.5}$$

of volume $V_0 = L_{0,x} \, L_{0,y} \, L_{0,z}$.

The DAM heavily relies on the appropriate definitions of porosities, which can be defined with respect to time and space. Porosity in time, ϕ_T, is defined as

$$\phi_T(\mathbf{x}, t) = \overline{\gamma}^s(\mathbf{x}, t) \qquad . \tag{4.6}$$

The integral in (4.2) can be identified with a duration T_f during which, in total, the point $\mathbf{x}$ was occupied with fluid, so that

$$T_f(\mathbf{x}, t) = \int_{\mathcal{T}(t)} \gamma(\mathbf{x}, t^*) \, dt^* = \int_{T_0} \gamma(\mathbf{x}, t^*) \, dt^* \tag{4.7}$$

with the second identity introducing the usual shorthand notation in which the dependency of the expression on t is no more visualized [112]. With these definitions, the temporal porosity is

$$\phi_T(\mathbf{x}, t) = \overline{\gamma}^s(\mathbf{x}, t) = \frac{T_f}{T_0} \qquad . \tag{4.8}$$

The same approach is also applied to averaging in space, defining spatial porosity as

$$\phi_V(\mathbf{x}, t) = \langle \gamma \rangle^s \qquad . \tag{4.9}$$

The integral in (4.4) is then replaced with

$$V_f(\mathbf{x}, t) = \int_{\mathcal{V}(\mathbf{x})} \gamma(\mathbf{x}^*, t) \, dV^* = \int_{V_0} \gamma(\mathbf{x}^*, t) \, dV^* \tag{4.10}$$

---

[1]Throughout this chapter, sets will be denoted with capital script letters and the size of a set with a capital roman letter.

again introducing a shorthand for the spatial integral. With these definitions, the spatial porosity is

$$\phi_V(\mathbf{x},t) = \langle\gamma\rangle^s(\mathbf{x},t) = \frac{V_f}{V_0} \quad . \tag{4.11}$$

Once both, averaging in time and averaging in space are defined, this can be combined to consecutive superficial time-space averaging

$$\langle\overline{\gamma}^s\rangle^s = \frac{1}{V_0}\int_{V_0}\frac{1}{T_0}\int_{T_0}\gamma(\mathbf{x}^*,t^*)\,dt^*\,dV^* \quad . \tag{4.12}$$

The related time-space porosity, the so-called total porosity, then is

$$\phi_{VT}(\mathbf{x},t) = \langle\overline{\gamma}^s\rangle^s \quad . \tag{4.13}$$

The averaging operations introduced above can be applied to any scalar quantity θ defined in the fluid phase by replacing γ in the above expressions with $\theta\gamma$ to define

$$\overline{\theta}^s = \frac{1}{T_0}\int_{T_0}\theta(\mathbf{x},t^*)\gamma(\mathbf{x},t^*)dt^* \quad , \tag{4.14}$$

$$\langle\theta\rangle^s = \frac{1}{V_0}\int_{V_0}\theta(\mathbf{x}^*,t)\gamma(\mathbf{x}^*,t)dV^* \quad , \tag{4.15}$$

$$\langle\overline{\theta}^s\rangle^s = \frac{1}{V_0}\int_{V_0}\frac{1}{T_0}\int_{T_0}\theta(\mathbf{x}^*,t^*)\gamma(\mathbf{x}^*,t^*)dt^*dV^* \quad , \tag{4.16}$$

where all quantities on the left-hand side of (4.14)-(4.16) depend on space and time. With the superficial average, integration in space, in time, or in space and time, is conducted regardless of the position of the disperse phase, i.e. regardless of $\gamma(\mathbf{x},t)$, and normalization is performed with the size of the entire integral, independent of γ, making this approach conceptually very simple.

Since the quantity θ is defined only in the fluid phase, it is also reasonable to only average over points in time and space, where the continuous phase is actually present. This so-called intrinsic time average of θ then is

$$\overline{\theta} = \frac{\int_{T_f}\theta(\mathbf{x},t^*)dt^*}{\int_{T_f}dt^*} \quad , \tag{4.17}$$

with $T_f(t) \subseteq T(t)$ the set of points in time, where fluid is present at the point $\mathbf{x}$. If the denominator in (4.17) vanishes, the nominator vanishes as well. In fact, the given point in space in this case did not see any fluid during the averaging interval in time so that no averaging of a fluid quantity can be performed.

Using the indicator γ, the integrals in (4.17) can be replaced by the integrals over the larger set $T(t)$, i.e.

$$\overline{\theta} = \frac{\int_{T_0}\theta(\mathbf{x},t^*)\gamma(\mathbf{x},t^*)dt^*}{\int_{T_0}\gamma(\mathbf{x},t^*)dt^*} \quad , \tag{4.18}$$

with the notation of (4.7). Multiplying nominator and denominator of this equation with $1/T_0$ and observing that the denominator in (4.18) equals T_f, it is obvious that superficial and intrinsic time average only differ by a spatially and temporally varying factor, the porosity in time, i.e.

$$\overline{\theta}^s = \phi_T \overline{\theta} \tag{4.19}$$

With the same precaution of non-vanishing denominator as for (4.17) and (4.18), the intrinsic spatial average is defined as

$$\langle \theta \rangle = \frac{\int_{V_0} \theta(\mathbf{x}^*, t)\gamma(\mathbf{x}^*, t)\mathrm{d}V^*}{\int_{V_0} \gamma(\mathbf{x}^*, t)\mathrm{d}V^*} \tag{4.20}$$

and the same considerations yield

$$\langle \theta \rangle^s = \phi_V \langle \theta \rangle \quad . \tag{4.21}$$

Considering the superficial average of the phase indicator γ, the above definitions yield

$$\phi_{VT} = \langle \overline{\gamma}^s \rangle^s = \langle \phi_T \rangle^s \quad , \tag{4.22}$$

where the second equality results from the definition (4.6). The essential observation now is that the phase indicator $\gamma(\mathbf{x}, t)$ does no more apply to the quantity

$$\phi_T(\mathbf{x}, t) = \frac{1}{T_0} \int_{\mathcal{T}(t)} \gamma(\mathbf{x}, t^*)\,\mathrm{d}t^* \tag{4.23}$$

itself. It suffices that $\gamma > 0$ at one instant t^* in $\mathcal{T}(t)$ to make $\phi_T > 0$ for all $t \in [t^* - T_0/2; t^* + T_0/2]$. The set of points in time and space where $\phi_T > 0$ hence is no more identified by γ but by a different indicator $\gamma_T(\mathbf{x}, t)$. The intrinsic average of ϕ_T in space, hence, is

$$\langle \phi_T \rangle = \frac{\int_{V_0} \phi_T(\mathbf{x}^*, t)\,\gamma_T(\mathbf{x}^*, t)\mathrm{d}V^*}{\int_{V_0} \gamma_T(\mathbf{x}^*, t)\mathrm{d}V^*} \quad , \tag{4.24}$$

provided the denominator is larger than zero. Denoting $\int_{V_0} \gamma_T(\mathbf{x}^*, t)\mathrm{d}V^* = V_m$ with $V_m \le V_0$ for the reason just described, and defining $\phi_{Vm} = V_m/V_0$ yields

$$\phi_{VT} = \langle \phi_T \rangle^s = \phi_{Vm}\langle \phi_T \rangle \quad , \tag{4.25}$$

with all these quantities depending on space and time.
Now suppose γ in (4.23) being multiplied by a fluid quantity θ. The temporal average $\overline{\theta}^s$ then has the same support in time as ϕ_T, namely γ_T, so that

$$\langle \overline{\theta}^s \rangle^s = \phi_{Vm}\langle \overline{\theta}^s \rangle \tag{4.26}$$

with ϕ_{Vm} defined as before. Applying furthermore relation (4.19) between both temporal averages yields the following relation for consecutive superficial time-space averaging

$$\langle \overline{\theta}^s \rangle^s = \frac{1}{V_0} \int_{V_0} \frac{1}{T_0} \int_{T_0} \theta(\mathbf{x}^*, t^*)\gamma(\mathbf{x}^*, t^*)\mathrm{d}t^*\mathrm{d}V^* = \phi_{Vm}\langle \phi_T \overline{\theta} \rangle \quad . \tag{4.27}$$

For consecutive intrinsic averaging in time and space, one obtains

$$\langle \overline{\theta} \rangle = \frac{1}{V_m} \int_{V_0} \frac{1}{T_f} \int_{T_0} \theta(\mathbf{x}^*, t^*)\gamma(\mathbf{x}^*, t^*)\mathrm{d}t^*\mathrm{d}V^* \quad . \tag{4.28}$$

4.3.2 Application of the DAM to the Navier-Stokes equations

The averaging operators defined in the previous section give rise to a modified Reynolds decomposition for intrinsic properties, since deviations in both time and space are possible. For convenience, we shift to index notation now and introduce the time-space Reynolds decomposition of the fluid quantities appearing in the Navier-Stokes equations (2.1), i.e. for fluid velocity and pressure

$$u_i = \overline{u}_i + u_i' \quad , \quad \overline{u}_i = \langle \overline{u}_i \rangle + \widetilde{\overline{u}}_i \quad , \tag{4.29a}$$

$$p = \overline{p} + p' \quad , \quad \overline{p} = \langle \overline{p} \rangle + \widetilde{\overline{p}} \quad . \tag{4.29b}$$

In (4.29), the prime indicates a deviation of the instantaneous value from its time-averaged value and the tilde indicates deviation of the time-averaged value from its spatially averaged value. No deviations in space are possible for the volume force f_i, because this quantity is constant in space by definition in the present numerical method as outlined in Chap. 2, thus giving

$$f_i(t) = \langle \overline{f}_i \rangle + f_i' \quad . \tag{4.30}$$

Here, the intrinsic averaging operator (4.28) is employed for the decomposition, using information in the fluid phase only. Therefore, $\mathbf{f}_{IBM}$ was neglected in this respect as it merely constitutes a numerical technique for imposing the boundary conditions on the particle surfaces. After substituting (4.29) and (4.30) in the Navier-Stokes equations (2.1), these can be rewritten as

$$\frac{\partial(\langle \overline{u}_i \rangle + \widetilde{\overline{u}}_i + u_i')}{\partial t} + \frac{\partial(\langle \overline{u}_i \rangle + \widetilde{\overline{u}}_i + u_i')(\langle \overline{u}_i \rangle + \widetilde{\overline{u}}_i + u_i')}{\partial x_j}$$
$$= (\langle \overline{f}_i \rangle + f_i') - \frac{1}{\rho_f}\frac{\partial(\langle \overline{p} \rangle + \widetilde{\overline{p}} + p')}{\partial x_i} + \frac{\partial}{\partial x_j}\left(\nu_f \frac{\partial(\langle \overline{u}_i \rangle + \widetilde{\overline{u}}_i + u_i')}{\partial x_j}\right) \quad . \tag{4.31}$$

Averaging of this equation in both time and space, i.e. double-averaging, can be performed using the superficial averaging operator (4.27). This averaging operator, however, requires the application of the chain rule, whenever derivatives are computed, because the fluid quantities as well as the clipping function indicating the different phases depend on space and time. This is typical for multiphase systems as outlined by [70], and results in additional source terms. These source terms were derived in [112] in form of the double-averaging theorems. Denoted for a scalar quantity $\theta(x_i, t)$, they read

$$\left\langle \phi_T \frac{\overline{\partial \theta}}{\partial t} \right\rangle = \frac{1}{\phi_{Vm}}\frac{\partial \phi_{Vm}\langle \phi_T \overline{\theta} \rangle}{\partial t} + \frac{1}{\phi_{Vm}V_0}\overline{\iint_{S_{int}} \theta v_i n_i \mathrm{d}S}^s \tag{4.32a}$$

and

$$\left\langle \phi_T \frac{\overline{\partial \theta}}{\partial x_i} \right\rangle = \frac{1}{\phi_{Vm}}\frac{\partial \phi_{Vm}\langle \phi_T \overline{\theta} \rangle}{\partial x_i} - \frac{1}{\phi_{Vm}V_0}\overline{\iint_{S_{int}} \theta n_i \mathrm{d}S}^s \quad , \tag{4.32b}$$

for derivatives in time and space, respectively. Here, S_{int} is the fluid-particle interface, v_i is the local velocity vector of the fluid-particle interface, and n_i is the unit vector normal to the particle surface and directed into the fluid. The next step is to introduce (4.32) into (4.31) and to apply

$$\overline{\overline{\theta}} = \overline{\theta}; \quad \langle\langle \theta \rangle\rangle = \langle \theta \rangle \quad , \tag{4.33a}$$

resulting in

$$\overline{\theta'} = 0; \qquad \left\langle \widetilde{\overline{\theta}} \right\rangle = 0 \quad . \tag{4.33b}$$

With the averaging operations defined in (4.17) and (4.20), relations (4.33a) are not satisfied automatically. For temporal averaging, the averaging interval in time must be sufficiently large, i.e. larger than the turbulent time scale. For spatial averaging, the averaging domain must be sufficiently large as well, i.e. larger than the grain scale [112]. Both requirements are met with parameters employed later on, so that (4.33a) can safely be applied to simplify the equations.

The steps described allow to transform Equation (4.31) into the double-averaged momentum balance [112]

$$
\underbrace{\frac{\partial \phi_{Vm} \langle \phi_T \overline{u}_i \rangle}{\partial t}}_{1} + \underbrace{\frac{\partial \phi_{Vm} \langle \phi_T \rangle \langle \overline{u}_i \rangle \langle \overline{u}_j \rangle}{\partial x_j}}_{2} = \underbrace{\phi_{Vm} \langle \phi_T f_i \rangle}_{3} - \underbrace{\frac{1}{\rho} \frac{\partial \phi_{Vm} \langle \phi_T \overline{p} \rangle}{\partial x_i}}_{4} -
$$

$$
\underbrace{\frac{\partial \phi_{Vm} \left\langle \phi_T \overline{u_i' u_j'} \right\rangle}{\partial x_j}}_{5} - \underbrace{\frac{\partial \phi_{Vm} \left\langle \phi_T \widetilde{\overline{u}}_i \widetilde{\overline{u}}_j \right\rangle}{\partial x_j}}_{6} + \underbrace{\frac{\partial}{\partial x_j} \left(\phi_{Vm} \left\langle \phi_T \nu \frac{\overline{\partial u_i}}{\partial x_j} \right\rangle \right)}_{7} -
$$

$$
\underbrace{\frac{\partial \phi_{Vm} \langle \phi_T \widetilde{\overline{u}}_i \rangle \langle \overline{u}_j \rangle}{\partial x_j}}_{8} - \underbrace{\frac{\partial \phi_{Vm} \langle \phi_T \widetilde{\overline{u}}_j \rangle \langle \overline{u}_i \rangle}{\partial x_j}}_{9} +
$$

$$
\underbrace{\frac{1}{\rho} \frac{1}{V_0} \overline{\iint_{S_{int}} p n_i \mathrm{d}S}^s}_{10} - \underbrace{\frac{1}{V_0} \overline{\iint_{S_{int}} \left(\nu \frac{\partial u_i}{\partial x_j} n_j \, \mathrm{d}S \right)}^s}_{11} \quad , \tag{4.34}
$$

which constitutes the basis of the analyses conducted in the following. With the same physical meaning as the terms of the Navier-Stokes equation, terms 1 and 2 in equation (4.34) represent local and convective accelerations, respectively. The third term represents the momentum supply by the driving volume force; the fourth term results from the pressure gradient, and term 7 contains the double-averaged viscous stresses. The fifth and sixth terms are additional contributions from turbulent and form-induced stresses that originate from the non-linear convection term in (4.31). So far, this procedure is similar to the derivation of additional turbulent stresses for the RANS-equations as presented, e.g., in [134], or for the phase-averaged Navier-Stokes equations as presented in [132]. In addition, the eighth and ninth terms represent momentum fluxes, i.e. stresses, due to potential spatial correlations between the local time porosity and time-averaged velocities and emerge from the non-linear convection term as well. The final two terms in (4.34), i.e. the tenth and eleventh, are interfacial terms arising from the double-averaging theorems for pressure and viscous drag (4.32b).

The surface terms introduced by the double-averaging theorems (4.32) add up to zero for the local acceleration (term 1) and the convective acceleration (term 2) due to the no-slip condition on S_{int} [112] so that they do not appear in (4.34). Observe that (4.34) is an equation for the fluid phase alone. Collision forces act between the particles directly when they touch and hence do not appear in (4.34). If the motion of particles is affected by collisions, though, this affects the surface terms and enters in (4.34) via this path. So far, the extent V_0 and T_0 of the local averaging in space and time, has not been specified. They can not be selected arbitrarily, without affecting the algebra leading to (4.34) as mentioned

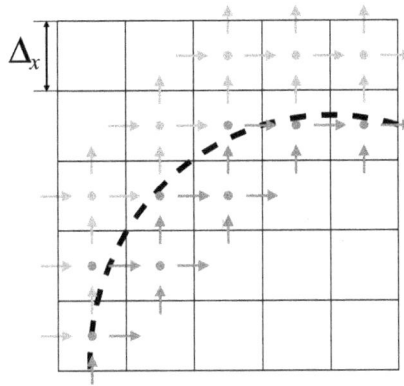

Figure 4.4: *Identification of the clipping function $\gamma(x_i, t)$ in the vicinity of a phase boundary (dashed line); arrows represent velocity vector components, red dots correspond to $\gamma(x_i, t) = 0$, green dots show $\gamma(x_i, t) = 1$.*

above. Nevertheless, there still is substantial freedom in how to choose these quantities at best for a given flow configuration, which will be discussed in Secs. 4.4.3 and 4.4.4 below.

4.4 Preparatory work

4.4.1 Numerical issues

The data from the DNS are available in form of a number of three-dimensional fields of velocity and pressure data at all points of the simulation grid together with all particle centers at the respective instants in time. The computational grid employed by the simulation code is staggered as described in Chap. 2 above and illustrated in Fig. 4.4. The step size of the grid is constant and the same in all directions. A staggered grid, however, is much too complicated for involved post-processing, so that all data is transfered to the cell-centered points of that grid. This is achieved by interpolation of the velocity components u_1, u_2, u_3 via

$$
\begin{aligned}
u_1^{i,j,k} &= \frac{1}{2}(u_1^{i-1/2,j,k} + u_1^{i+1/2,j,k}) \\
u_2^{i,j,k} &= \frac{1}{2}(u_2^{i,j-1/2,k} + u_2^{i,j+1/2,k}) \\
u_3^{i,j,k} &= \frac{1}{2}(u_2^{i,j,k-1/2} + u_2^{i,j,k+1/2})
\end{aligned}
\tag{4.35}
$$

where integer superscripts denote the index of the center of a cell and half indices the respective centers of cell faces. Pressure data does not require interpolation as these are already cell centered.

The next step is to determine the clipping function γ, also in form of cell-center values. In the space-discretized form, the clipping function is located at the cell center and must become zero, whenever fluid information from the inner of a solid object was used for the interpolation routine (4.35) (red arrows in Fig. 4.4). Therefore, every cell with the position vector $\mathbf{x}$ was evaluated by

$$
\begin{aligned}
\boldsymbol{\xi}_x &= min((\mathbf{x} - 0.5\Delta_x\mathbf{e}_1 - \mathbf{x}_p), (\mathbf{x} + 0.5\Delta_x\mathbf{e}_1 - \mathbf{x}_p)) \\
\boldsymbol{\xi}_y &= min((\mathbf{x} - 0.5\Delta_y\mathbf{e}_2 - \mathbf{x}_p), (\mathbf{x} + 0.5\Delta_y\mathbf{e}_2 - \mathbf{x}_p)) \\
\boldsymbol{\xi}_z &= min((\mathbf{x} - 0.5\Delta_z\mathbf{e}_3 - \mathbf{x}_p), (\mathbf{x} + 0.5\Delta_z\mathbf{e}_3 - \mathbf{x}_p))
\end{aligned}
\tag{4.36}
$$

to indicate the liquid and the solid phases, with $\mathbf{e}_i$ the unit vector in the *ith* direction. Consequently, γ in its space-discretized form becomes

$$
\gamma^{i,j,k} = \begin{cases} 1 & , min(|\boldsymbol{\xi}_x|, |\boldsymbol{\xi}_y|, |\boldsymbol{\xi}_z|) > R_p \\ 0 & , min(|\boldsymbol{\xi}_x|, |\boldsymbol{\xi}_y|, |\boldsymbol{\xi}_z|) < R_p \end{cases} .
\tag{4.37}
$$

This condition is illustrated in Fig. 4.4, where $\gamma(x_i, t)$ becomes equal to zero for the red dots.

This in fact increases the solid domain by a tiny amount, but was done intentionally to discard all numerical issues which might arise from the IBM approach in the close vicinity of the interfaces. The related volume is so small that it is negligible in the averaging process applied later on.

The numerical evaluation of averages requires a few remarks, as later on derivatives of the averaged quantities have to be computed to evaluate the individual terms of (4.34). This is discussed now by means of the average in x-direction. Similar considerations apply to averages in y, z, and t. For reasons which become clear immediately, the averaging extent $L_{0,x}$ is chosen to be an integer multiple of the step size of the DNS grid Δ_x, and an integer fraction of the domain size, L_x, such that $KL_{0,x} = L_x$ for some integer K. With (4.15), (4.28), etc., average quantities are defined as so-called sliding averages, yielding a continuous function in space. To obtain a coarser representation, the averaged quantities are evaluated only at certain points $X_i = L_{0,x}/2 + iL_{0,x}, i = 1, ..., K$. All spatially averaged data are then stored on this coarser grid. Due to the choice of $L_{0,x}$, this grid is regular and the averaged data fulfill the same periodic boundary conditions as the original DNS data. Derivatives are then evaluated by applying central finite-difference formulas of second order on this coarser grid. For the present configuration, data are not periodic in y and t, so that one-sided differences are needed for this purpose, but the later choices will not require these derivatives.

Remark: The reader might recognize that the restriction of the average data to the coarser grid just described may also be interpreted in a finite-volume sense. The DNS data obtained with a finite-volume code represent averages of the unknowns over cells with step size Δ_x. Merging these into larger cells of size $L_{0,x}$ results in a finite-volume representation with the larger cells. The definition of the spatial average based on the sliding volume $\mathcal{V}(x)$, when evaluated at the center of the large cell X_i, yields exactly the same result. Derivatives would then be evaluated in a finite-volume sense. Central finite volume schemes of second order on a regular grid, however, are identical to the corresponding, finite-difference schemes under these conditions [129].

4.4.2 Minimal schematic example

This subsection aims at illustrating the various porosities defined in Sec. 4.3.1 and the numerical evaluation of averages described in Sec. 4.4.1 by a minimal schematic example. Suppose that $\Delta_x = \Delta_y = \Delta_t = 1$ and consider a two-dimensional quadratic domain $[0; 2] \times$

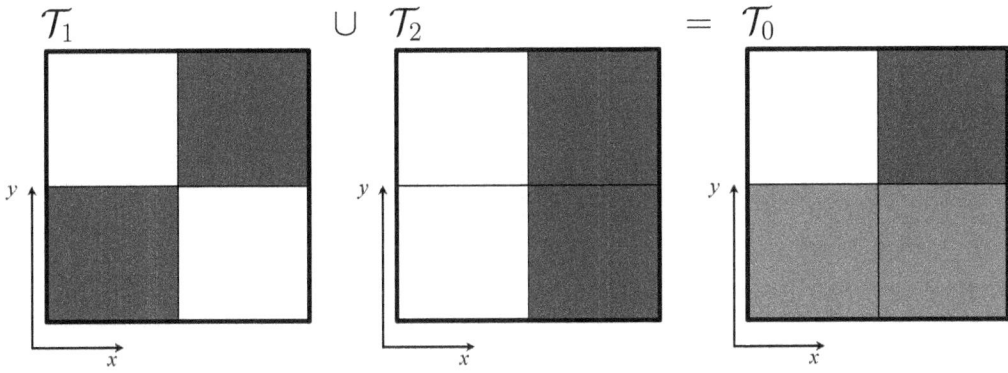

Figure 4.5: *Illustration of the double-averaging operator by a minimal schematic example. Blank area is occupied by fluid, gray area with stripe pattern of solid lines is occupied by solid. The left scheme represents the situation during the first interval in time $\mathcal{T}_1$, the middle picture during the second interval $\mathcal{T}_2$. The right scheme represents the average over the interval $\mathcal{T}_0 = \mathcal{T}_1 + \mathcal{T}_2$. The light grey area with a stripe pattern of dashed lines was occupied by solid and fluid during $\mathcal{T}_0$.*

$[0; 2]$ with periodic conditions in both directions. Suppose the raw data as illustrated in Fig. 4.5. During time interval $\mathcal{T}_1 = [0, 1[$, fluid covers the upper left and lower right unit square, while the other two squares are filled with solid. During the time interval $\mathcal{T}_2 = [1, 2]$, the solid originally in the lower left square is located in the lower right square. The marker function $\gamma(\mathbf{x}, t)$ is hence discontinuous not only in space but also in time for this example.

The right graph in Fig. 4.5 illustrates the superficial temporal average of γ, the porosity ϕ_T. The upper left square was occupied with fluid throughout, i.e. $T_f = 2, \phi_T = 1$, the dark gray area in the upper right square was not visited by fluid at all, hence $T_f = 0, \phi_T = 0$, and the light gray area has $T_f = 1, \phi_T = 1/2$. As a second step, averaging in space is performed with $L_{0,x} = 1$ and $L_{0,y} = 2$. According to the remarks in the previous section, averages are evaluated on a coarser grid, with spacing $L_{0,x}$ and $L_{0,y}$ in x and y, respectively. This leads to the data points depicted in Fig. 4.6. The values of the different porosities, $\phi_{VT}, \phi_{Vm}, \langle \phi_T \rangle$, can be computed by hand for this example, applying the respective definitions, so as to gain a better understanding of these quantities. This is helpful as they appear as space- and time-dependent factors in the double-averaged momentum balance (4.34).

4.4.3   Selection of averaging time

The scenarios *HP* and *LP* had run until stationary and uniform flow conditions were reached, before data recording for statistical analysis was started. The simulation duration of this second stage needs to be long enough to provide statistically stable averaged quantities when evaluating the terms in the double-averaged momentum equations for steady uniform flow. To determine an appropriate duration, first- and second-order velocity statistics, $\overline{u}$, $\overline{u'u'}$, and $\overline{u'v'}$, were computed as a function of the averaging time T_0. For sufficiently large T_0, the statistical parameters approach their expected constant values and thus the results become independent of T_0. This condition was tested for 3 different wall-normal coordinates and 6 spanwise coordinates at $x = 0$, which gives a total of 18 test locations distributed evenly over the cross section. This analysis is illustrated in Fig. 4.7 for two selected spanwise coordinates of scenario *HP*: i) $z = 2H$, located at the stable ridge which extends over the full

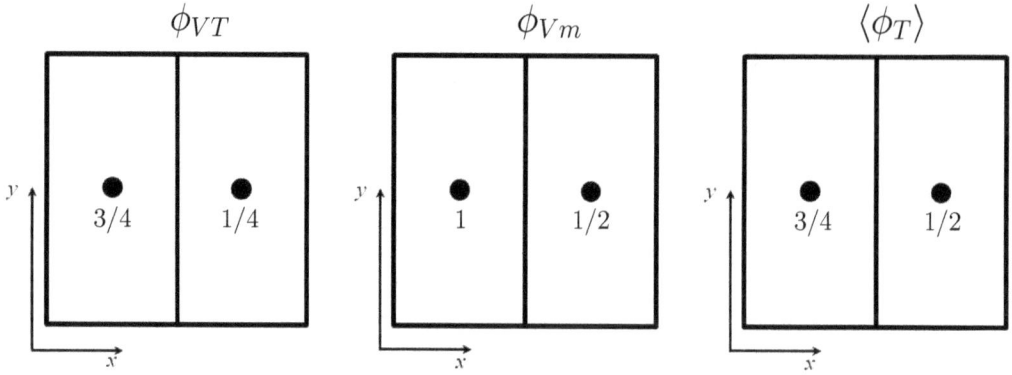

Figure 4.6: Porosities calculated from the schematic illustrated in Fig. 4.5 using the discrete averaging operators discussed in the text.

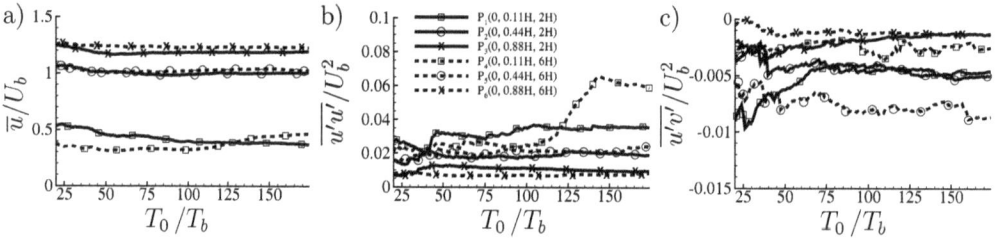

Figure 4.7:   *Time averaged quantities for scenario* HP *at the cross section* $x = 0$ *for three selected wall-normal coordinates and two selected spanwise coordinates: a)* $\overline{u}/U_b$, *b)* $\overline{u'u'}/U_b^2$, *and c)* $\overline{u'v'}/U_b^2$.

streamwise length of the simulation domain; and ii) $z = 6H$ at the unstable particle cluster that moves slowly over the fixed bed in streamwise direction.

While the statistics computed for the locations above the stable ridge at $z = 2H$ are well converged for all coordinates and statistical quantities explored, the situation is different for the locations above the unstable ridge at $z = 6H$. At point P_5 and P_6 in the outer flow region well above moving particles, the first-order velocity statistics approach constant values (Fig. 4.7a). In the near-wall region, the time-averaged velocity $\overline{u}$ starts to increase beyond $T_0 > 100T_b$. This trend becomes even more evident considering the second-order statistics in Fig. 4.7b, where even in the outer flow region changes are noticeable. The time evolution of the particle clusters inspected by visualization (not shown here) reveal that the position $P_4 = (0, 0.11H, 6H)$ is significantly influenced by the slowly moving cluster until $T_0 = 95T_b$. As soon as the cluster has passed, the hiding and shading mechanisms slowing down the fluid in the vicinity of the particles break down and the local fluid velocity as well as the turbulent fluctuations become stronger. This feature is most pronounced for position P_4, while for the other points investigated no such drastic changes are visible.

These tests illustrate that different regions of the simulation domain can exhibit different statistical behavior, influenced by either particle clusters or by "clear-water" flow, if particles are close to their critical value of the Shields parameter. Following the arguments of [112], the averaging time must well exceed the turbulent time scale T_b, but still has to be much shorter than the morphological time scale of the moving particle clusters. For the scenario with light particles (*LP*), the homogeneous distribution of the particles leads to much faster

a)

b)

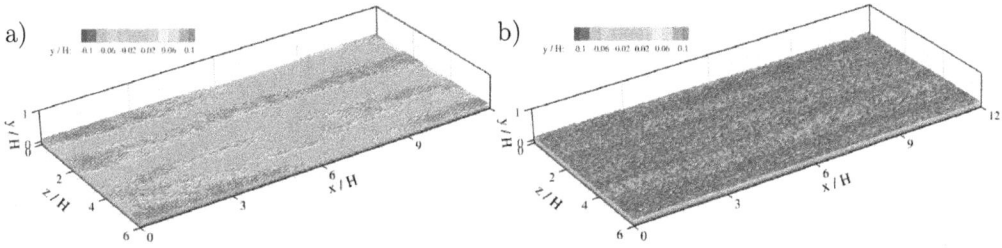

Figure 4.8: *Iso-surfaces of time-averaged volume porosity ϕ_T colored by the wall-normal elevation. The value of the iso-surface corresponds to the peak of $\langle \phi_T \rangle$ at $y = 0.1H$ illustrated in Fig. 4.11 below: a) scenario HP, b) scenario LP.*

a) Quadratic extent		b) Streamwise extent	
Option	Extent	Option	Extent
$V_{q,1}$	$1D \times \Delta_y \times 1D$	$V_{x,1}$	$1D \times \Delta_y \times 2D$
$V_{q,2}$	$2D \times \Delta_y \times 2D$	$V_{x,2}$	$2D \times \Delta_y \times 2D$
$V_{q,4}$	$4D \times \Delta_y \times 4D$	$V_{x,4}$	$4D \times \Delta_y \times 2D$
$V_{q,6}$	$6D \times \Delta_y \times 6D$	$V_{x,9}$	$1H \times \Delta_y \times 2D$
$V_{q,9}$	$9D \times \Delta_y \times 9D$	$V_{x,18}$	$2H \times \Delta_y \times 2D$
$V_{q,18}$	$18D \times \Delta_y \times 18D$	$V_{x,36}$	$4H \times \Delta_y \times 2D$
		$V_{x,72}$	$6H \times \Delta_y \times 2D$
		$V_{x,108}$	$12H \times \Delta_y \times 2D$

Table 4.4: *Investigated variations of the extent of the averaging domain.*

statistical convergence compared to the *HP* case. Hence, a shorter duration of averaging was used for the *LP* case (Tab. 4.3). Based on the investigations reported, it was decided to use the total simulation duration as the averaging time for both scenarios to assure that statistical convergence is maximized. The time-averaged, three-dimensional local porosity of the two scenarios is illustrated in Fig. 4.8.

4.4.4   Selection of spatial averaging domain

The spatial averaging volume V_0 needs to compromise between two requirements. On the one hand, it needs to be fine enough to resolve spatial heterogeneity. On the other hand, it must be large enough to average out the grain scale. Due to the boundary conditions employed, an open-channel flow is anisotropic and inhomogeneous in wall-normal direction and profiles along this direction exhibit strong gradients. To properly describe these gradients, a fine resolution was chosen so that the wall-normal extent of the averaging domain is equal to the cell size of the numerical grid in vertical direction, i.e. $L_{0,y} = \Delta_y = D/22.2 = 0.005H$, which is the finest resolution possible, even satisfying the prerequisite of a DNS to resolve the smallest scales of the flow and actually corresponds to no averaging in post-processing.

   For the horizontal directions, a distribution of the porosity close to homogeneous conditions can be reported in streamwise direction for sufficient averaging time (Fig. 4.8), while the present scenarios reveal heterogeneity in spanwise direction due to the particle clusters described in Sec. 4.2.1. This heterogeneity is more pronounced for scenario *HP* as the

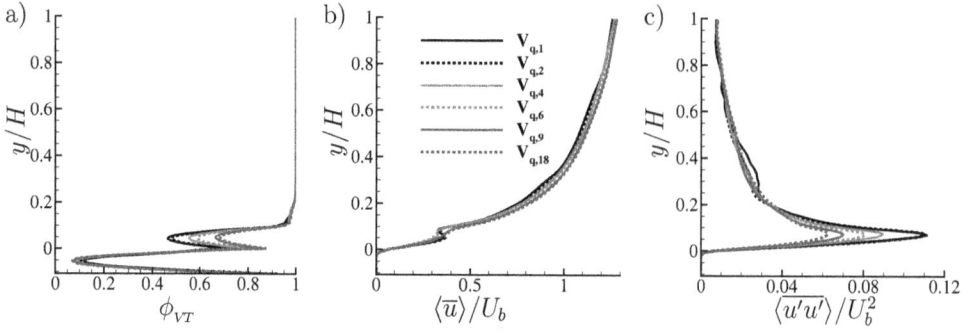

Figure 4.9: *Wall-normal profiles for different choices of a quadratic averaging domain V_q for scenario HP: a) total porosity, b) fluid velocity (streamwise component), and c) fluid fluctuations (streamwise component). The origin of the local coordinate system of the averaging domain corresponds to P_4 addressed in Fig. 4.7.*

roughness elements are constituted of mostly resting particles, while the situation is close to homogeneous in scenario *LP*. The spanwise extent of the averaging domain needs to account for this heterogeneity. Since the spanwise spacing of the resting particle clusters of scenario *HP* is $2H$ on average, an extent of the averaging volume larger than $9D$ in this direction would average out the heterogeneity of the particle pattern.

The impact of the size of V_0 on the averaging procedure described in Sec. 4.3.1 was explored by calculating wall-normal profiles for different choices of the horizontal extent of V_0. This was done for the spanwise positions $z/H = 2, 4, 6$ using an averaging volume with a quadratic horizontal extent $V_{q,m}$ with m a natural number identifying the size in multiples of D, as detailed in Tab. 4.4a. Fig. 4.9 shows the wall-normal profiles for different choices of $V_{q,m}$ for P_4 at $z/H = 6$, because it was found to be the most delicate point in Sec. 4.4.3. In order to describe steady-state conditions, this effect can be compensated by a suitable size of the averaging domain. The six options illustrated can be grouped into three pairs: i) a local averaging domain with a horizontal extent on the order of the size of the particle diameter D, ii) an intermediate extent larger than D but smaller than the channel height H, and iii) a large size where the horizontal extent is a multiple of H. The local domains show the lowest porosity and the strongest fluctuations in the near-wall region, while the large domains show the highest porosity and the lowest fluctuations (see Fig. 4.9a and 4.9c). In other words, the larger averaging domains average over ridges and troughs. Hence, an averaging domain with a local extent of less than four particle diameters is necessary to properly resolve the heterogeneity in spanwise direction. Option $V_{q,1}$, however, shows a wavy profile of the turbulent fluctuations in the outer flow (Fig. 4.9c) indicating that this averaging volume is too small to provide adequate statistics for the present analysis On the other hand, this effect can also be remedied by only increasing the streamwise extent of the averaging domain as shown below. On the basis of these results, a spanwise averaging over $2D$ was selected.

This being fixed, the streamwise extent of the averaging volume was explored with the investigated volumes listed in Tab. 4.4b. Since homogeneous conditions are present in streamwise direction, option $V_{x,108}$ covering the full streamwise extent of the computational domain $L_x = 108D$ contains the largest number of samples and hence should give the best results for the computed statistics. It can, therefore, be taken as a reference to compare, if the options with a smaller extent collapse with the profile of $V_{x,108}$.

Since P_4 was identified to be the most critical point in the domain in terms of statistical

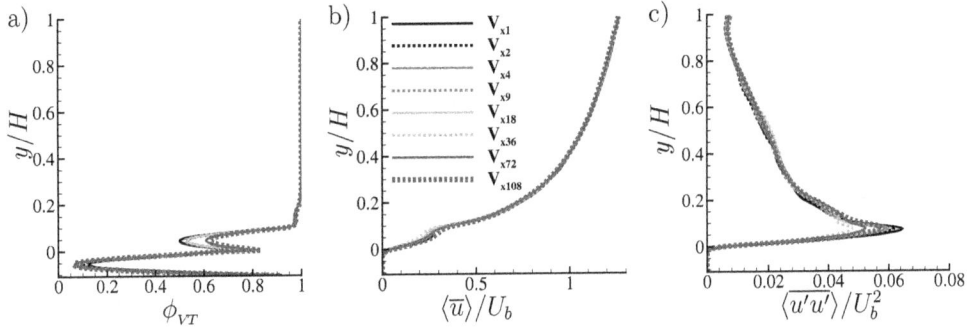

Figure 4.10: *Wall-normal profiles for different choices of a quadratic averaging domain V_x for scenario HP: a) total porosity, b) fluid velocity (streamwise component), and c) fluid fluctuations (streamwise component). The origin of the local coordinate system of the averaging domain corresponds to P_4 addressed in Fig. 4.7.*

convergence, the wall-normal profiles of this analysis are shown for P_4 in Fig. 4.10. The statistics of the total porosity and the averaged streamwise fluid velocity converge for an averaging volume covering more than $72D = 0.5L_x$. The velocity variance is converged in the outer flow (Fig. 4.10c), but slight deviations of $\leq 15\%$ remain in the near-wall regions of the profile of the velocity variance at P_4, when comparing the results of $V_{x,72}$ with the reference volume $V_{x,108}$. This is the inevitable compromise one has to take into account, if the averaging time is chosen by the criteria mentioned in Sec. 4.4.3. Differences were substantially smaller at all other points

According to [112] a suitable averaging domain must be chosen such that its extent is larger than the roughness scales, i.e. the particle diameter, and smaller than large scale morphological features such as the spacing of the ridges reported for case *HP*. Therefore, two averaging domains are used in the main analysis to compare the different scales observed. First, a local averaging domain of size

$$V_{0,2} = L_x \, \Delta_y \, 2D \qquad (4.38)$$

is chosen to investigate local heterogeneity in spanwise direction. In the following, this averaging procedure will be referred to as "local averaging". Second, the total spanwise extent was used averaging over volumes of size

$$V_{0,1} = L_x \, \Delta_y \, L_z \qquad (4.39)$$

to address the large scale effects. This averaging procedure will be referred to as "global averaging".

Note that with both choices derivatives in streamwise direction vanish due to the periodic boundary conditions in x-direction. This yields two-dimensional distributions for the local averaging. Furthermore, applying the global averaging volume $V_{0,1}$ to the averaging operators defined in Sec. 4.3.1 above, the derivatives in z-direction vanish for the same reason giving one-dimensional profiles of the same kind as presented in [157].

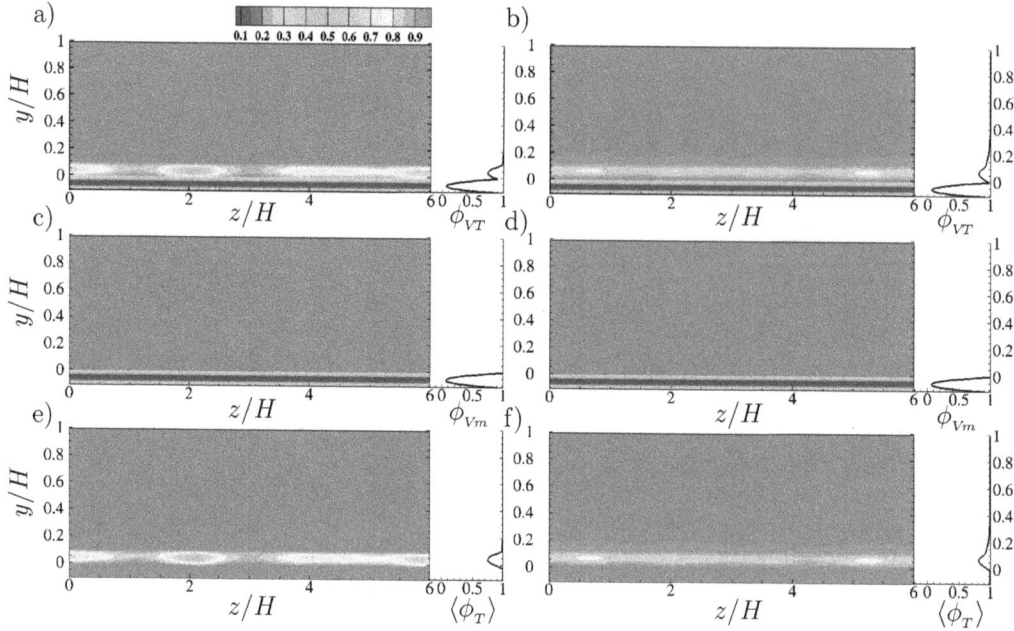

Figure 4.11: *Spatial distribution of the double averaged porosities for the scenarios* HP *(left) and*
LP *(right). Contour plots are computed using a local average in z of volume $V_{0,2}$ and wall-normal*
profiles using the global average employing $V_{0,1}$. a) Total porosity ϕ_{VT} for scenario HP*, b) ϕ_{VT}*
for scenario LP*, c) space porosity ϕ_{Vm} for scenario* HP*, d) ϕ_{Vm} for scenario* LP*, e) the spatially*
averaged time porosity $\langle\phi_T\rangle$ for scenario HP *f) $\langle\phi_T\rangle$ for scenario* LP*. Horizontal and vertical axis*
are not to scale.

4.5 Moment fluxes and balance within the double - averaged framework

4.5.1 Porosities and double-averaged velocities

As the DAM aims to account for spatial heterogeneity of the flow by introducing the porosity in the averaged momentum balance, the distribution of the total porosity ϕ_{VT} governs
the local magnitude of the double-averaged quantities. The spatial distribution of these
quantities is shown in Fig. 4.11–4.23. Throughout the paper, these figures are organized as
follows. Quantities related to scenario *HP* are placed on the left hand side, while the results
of scenario *LP* are on the right hand side to allow easy comparison. The spatial distribution
of each quantity is displayed as a two-dimensional plot of the locally averaged quantities
with the color scale always the same for all graphs assembled in one figure. Observe that
horizontal and vertical axis in the two-dimensional plots are not to scale to better display
the results. A one-dimensional plot represents the global average using $V_{0,1}$, again with the
range being the same in all graphs of a figure.

According to equation (4.25), the total time-space porosity ϕ_{VT} can be decomposed into the
product of the spatial porosity ϕ_{Vm} and the spatially averaged temporal porosity $\langle\phi_T\rangle$. In
this regard, $1-\phi_{Vm}$ represents the part of the domain that has not been visited at all by fluid
during the averaging time T_0, while $\langle\phi_T\rangle$ is a measure for the porosity of the mobile granular

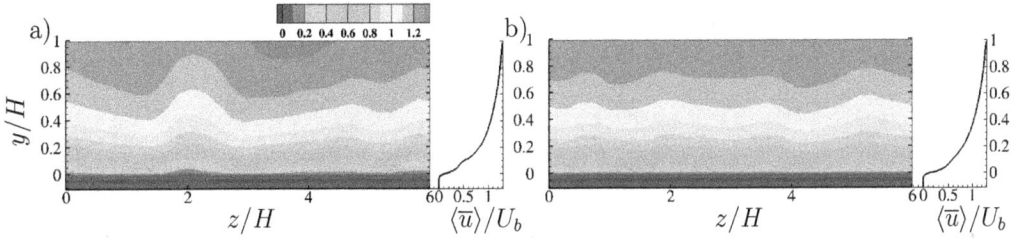

Figure 4.12: *Spatial distribution of the time-space averaged velocity $\langle \overline{u} \rangle / U_b$. a) Scenario HP, b) scenario LP. Horizontal and vertical axis are not to scale.*

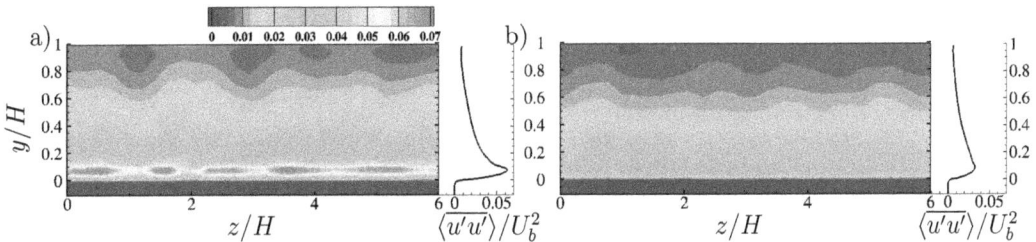

Figure 4.13: *Spatial distribution of the time-space averaged fluctuations $\langle \overline{u'u'} \rangle / U_b^2$. a) Scenario HP, b) scenario LP. Horizontal and vertical axis are not to scale.*

bed. Fig. 4.11 reveals that for $y > 0$, i.e. above the layer of fixed particles representing the rough wall, $\phi_{Vm} = 1$ for both cases *HP* and *LP*. This stems from the fact that, although the ridges are stable patterns, the individual particles constituting the ridges move, at least from time to time, so that each point above the fixed bed is met by fluid during the averaging time. Hence, the ridges observed in scenario *HP* are related to the spatial distribution of $\langle \phi_T \rangle$ above and can be discussed in the respective graphs. Low values are obtained for the region $0 < y < 0.1H$, in particular at $z/H = 2$, $z/H = 4$, and $z/H = 6$ with the one-dimensional profile of the global average peaking at the elevation of $y = 0.056H = 0.5D$. At these locations, the low porosity decelerates the fluid over the whole channel height and introduces a kink in the globally averaged velocity profile at $y = 0.11H = 1D$ (Fig. 4.12a). The porosity of scenario *LP* shows a distribution closer to homogeneous conditions. Nevertheless, four quite regularly spaced maxima of $\langle \phi_T \rangle$ and ϕ_{VT} can be discovered at $z/h \approx 0.6, 2.1, 3.8, 5.3$ impacting on the average $\langle \overline{u} \rangle$ displayed in Fig. 4.12b. The period of this pattern in z is $1.5H$ and hence shorter as for scenario *HP*, but still regular. This is unexpected and could not be detected from animations or snapshots like shown in Fig. 4.3, thus demonstrating the usefulness of the method. The kink in the one-dimensional profile reported for case *HP* is not present for case *LP*. Using similar thresholds as suggested by [12], one can identify the bed-load layer of the granular bed with a threshold of $\langle \phi_T \rangle < 0.99$ in the one-dimensional profile. This was found to be $0 < y < 0.18H$ for case *HP* and $0 < y < 0.35H$ for case *LP*.

4.5.2   Momentum fluxes

As expected, the distribution of the porosities observed for scenario *HP* in Fig. 4.11 and 4.12 is reflected in the four components of the turbulent flux. Among the three components investigated, the turbulent flux of the streamwise fluctuations $\langle \overline{u'u'} \rangle$ is larger (Fig. 4.13), compared to the fluctuations of the wall-normal velocity $\langle \overline{v'v'} \rangle$, and the spanwise velocity

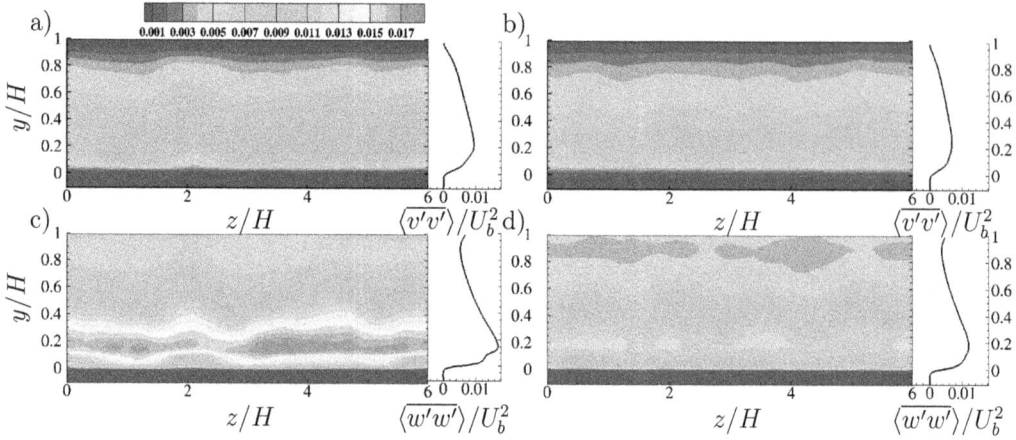

Figure 4.14: *Spatial distribution of the time-space averaged wall-normal and spanwise fluctuations. a) $\langle \overline{v'v'}\rangle/U_b^2$ for scenario HP, b) $\langle \overline{v'v'}\rangle/U_b^2$ for scenario LP, c) $\langle \overline{w'w'}\rangle/U_b^2$ for scenario HP, d) $\langle \overline{w'w'}\rangle/U_b^2$ for scenario LP. Horizontal and vertical axis are not to scale.*

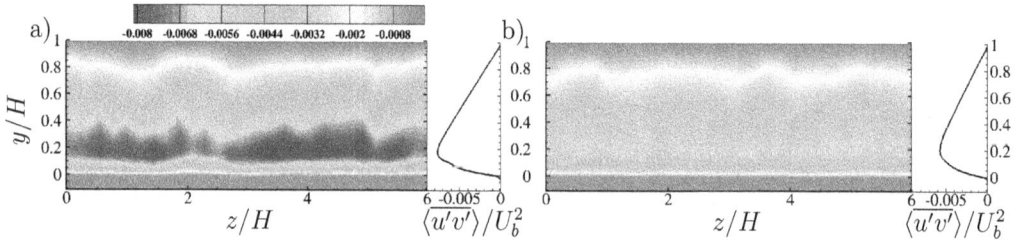

Figure 4.15: *Spatial distribution of the time-space averaged fluctuations $\langle \overline{u'v'}\rangle/U_b^2$. a) Scenario HP, b) scenario LP. Horizontal and vertical axis are not to scale.*

$\langle \overline{w'w'}\rangle$ (Fig. 4.14). Close to the free surface at the top, kinetic energy is transferred from the v-component to the two other velocity components, which is a well-known fact [110] and will not be discussed further here.

Starting with scenario *HP*, it is noticed that $\langle \overline{u'u'}\rangle$ exhibits maxima on both sides of the ridges. This is visible at elevations $y \approx 0.07H$ but even much higher above by the distortion of the level lines. The global average of the Reynolds shear stress $\langle \overline{u'v'}\rangle$ displayed in the one-dimensional profiles of Fig. 4.15a peaks directly above the layer of mobile particles at $y \approx 1.5D = 0.167H$. A kink at $y \approx 1D = 0.11H$ in the one-dimensional profile is observed for $\langle \overline{w'w'}\rangle$ and $\langle \overline{u'v'}\rangle$ and is caused by the stresses on top of the resting particles.

The results for case *HP* are now compared to those of case *LP*. Although the number of mobile particles introduced is the same in both simulations, the turbulent fluxes shown in the right parts of Figs. 4.13-4.15 differ quantitatively from those in case *HP*. Their globally averaged magnitude is smaller by up to 50 %. This is due to the fact that the momentum exchange of the light particles with the rough bottom by collision and contact is smaller resulting in a lower relative velocity. The effect can be quantified by the particle Reynolds number based on the globally averaged slip velocity $\langle \overline{u_p}\rangle - \langle \overline{u}\rangle$, which is larger for case *HP* compared to case *LP*, and exhibits a local maximum around $y_p = 1.32D$ not present for *LP* [151]. For scenario *LP*, the local distribution of the turbulent fluxes is substantially more homogeneous, and the one-dimensional profiles do not show the distinct kinks reported for

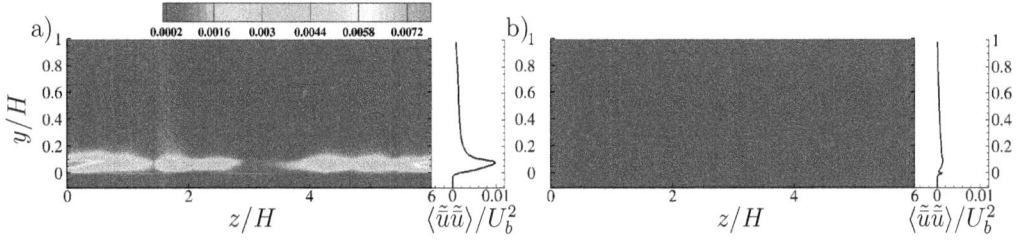

Figure 4.16: *Spatial distribution of the time-space averaged spatial stress $\langle \tilde{\bar{u}}\tilde{\bar{u}} \rangle / U_b^2$. a) Scenario HP, b) scenario LP. Horizontal and vertical axis are not to scale.*

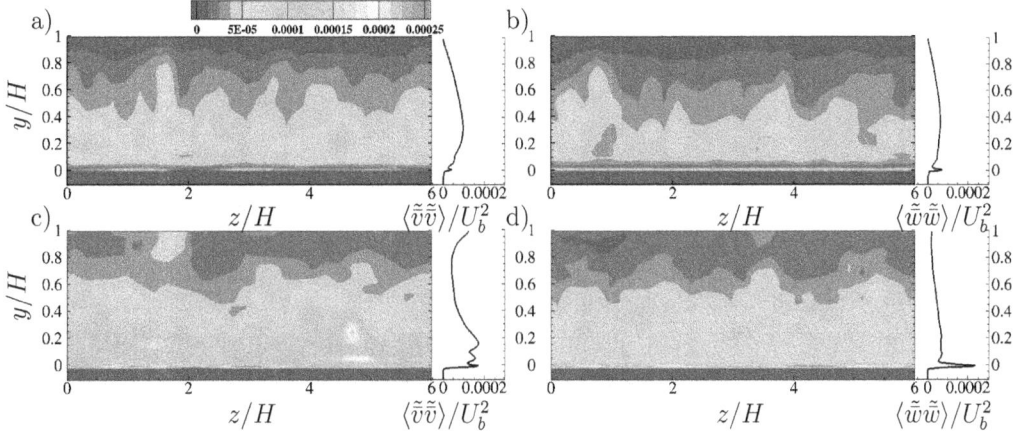

Figure 4.17: *Spatial distribution of the time-space averaged form-induced fluctuations. a) $\langle \tilde{\bar{v}}\tilde{\bar{v}} \rangle / U_b^2$ for scenario HP, b) $\langle \tilde{\bar{v}}\tilde{\bar{v}} \rangle / U_b^2$ for scenario LP, c) $\langle \tilde{\bar{w}}\tilde{\bar{w}} \rangle / U_b^2$ for scenario HP, d) $\langle \tilde{\bar{w}}\tilde{\bar{w}} \rangle / U_b^2$ for scenario LP. Horizontal and vertical axis are not to scale.*

$\langle \overline{w'w'} \rangle$ and $\langle \overline{u'v'} \rangle$ in the *HP* scenario. Slight distortions in the level lines of $\langle \overline{u'u'} \rangle$ are visible, though, and localized near the points where $\langle \phi_T \rangle$ has locally larger values at the bottom.

The locally averaged form-induced momentum fluxes $\langle \tilde{\bar{u}}_i \tilde{\bar{u}}_j \rangle$ illustrate the effects of heterogeneity of the particle distribution for the two scenarios. For case *HP*, the streamwise component $\langle \tilde{\bar{u}}\tilde{\bar{u}} \rangle$ shows large values in the near-wall region (Fig. 4.16a), which are of the same order of magnitude as reported for the turbulent momentum fluxes $\langle \overline{u'u'} \rangle$. In the outer flow, $\langle \tilde{\bar{u}}\tilde{\bar{u}} \rangle$ becomes small. Compared to the maximum of $\langle \tilde{\bar{u}}\tilde{\bar{u}} \rangle$, the other two normal components, $\langle \tilde{\bar{v}}\tilde{\bar{v}} \rangle$ and $\langle \tilde{\bar{w}}\tilde{\bar{w}} \rangle$ are two orders of magnitude smaller (Fig. 4.17a and 4.17c). The form-induced shear stress $\langle \tilde{\bar{u}}\tilde{\bar{v}} \rangle$ of case *HP* is displayed in Fig. 4.18a. Similar values to those of $\langle \tilde{\bar{u}}\tilde{\bar{u}} \rangle$ are obtained. In particular around $z \approx 2H$ it can be observed how this quantity reflects the secondary flow induced by the lower porosity, a feature which will be discussed in more detail below. At $z \approx 2H$, $\tilde{\bar{u}} > 0$ as illustrated by Fig. 4.12a above since spatial averaging is only local. Furthermore, $\tilde{\bar{v}} > 0$ at this location, so that $\langle \tilde{\bar{u}}\tilde{\bar{v}} \rangle > 0$. On both sides, a negative region is generated by an average downward current. A similar pattern is observed for $z \approx 6H$. Note that $\langle \tilde{\bar{u}}\tilde{\bar{v}} \rangle$ based on global averaging is about two orders of magnitude smaller.

For case *LP*, only small values of $\langle \tilde{\bar{u}}\tilde{\bar{u}} \rangle$ are obtained in the entire domain (Fig. 4.16b). For the global average, a strong peak on top of the fixed bed is visible at $y = 0$ for $\langle \tilde{\bar{w}}\tilde{\bar{w}} \rangle$ (Fig. 4.17d) which is not present in scenario *HP*. This indicates a sharp transition between the flow in and around the fixed bed to the outer flow. In scenario *HP*, this transition is not as distinct, be-

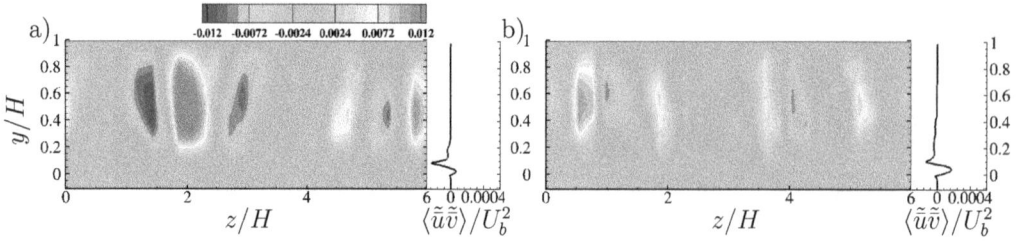

Figure 4.18:   *Spatial distribution of the time-space averaged spatial stress $\langle \tilde{\bar{u}}\tilde{\bar{v}} \rangle / U_b^2$. a) Scenario HP, b) scenario LP. Horizontal and vertical axis are not to scale.*

cause here the particles are temporally resting on top of the fixed bed slowing down the fluid.

4.5.3   Rates of change of momentum

After investigating the turbulent fluxes, this subsection addresses the full double-averaged momentum equation (4.34). For modelers it is highly relevant to know which of these terms is large, hence requiring a sound model, and which of the terms can be neglected. This information can be obtained from the present DNS data by computing each individual term in this equation and evaluating its magnitude. As a grain of salt, however, this is not possible for term 10 and term 11, which contain integrals over the phase boundary. The IBM yields a smoothed velocity field and employs a computational trick to evaluate the integrals over the viscous forces and the pressure force [148], so that local forces are not available. Terms 10 and 11 hence can not be computed *a posteriori* for dense packings. Term 1 in this equation drops out, since the averaging time was chosen equal to the full duration of the signal, so that the temporal derivative in (4.34) vanishes. Hence, terms 2 to 9 in (4.34) were computed employing the local spatial averaging according to (4.38) and the discretization described in Sec. 4.4.1. Since this involves averaging over L_x in streamwise direction, the terms containing derivations in x_1 equate to zero and are hence omitted. This concerns all terms with $j = 1$. Tab. 4.5 and 4.6 assemble the maxima of the absolute values of all computed non-vanishing terms for scenario *HP* and *LP*, respectively, to provide an impression of their magnitude and their relevance for the momentum balance. To maintain the non-dimensional notation, the normalization by U_b^2/H is used throughout but omitted in the following for brevity.

The momentum exchange between the outer flow and the fixed and mobile sediment is governed by the off-diagonal terms $i = 1$ and $j = 2$. Indeed, these terms are mostly large, as visible from Tab. 4.5 and 4.6 and will now be discussed in more detail by means of two-dimensional contour plots reported in Figs. 4.19 – 4.23. In addition to this, significant convective and form-induced stresses are reported for $i = 1$ and $j = 3$ in Tab. 4.5 and 4.6 indicating a transverse transfer of momentum in the channel. This issue will be addressed in detail in Sec. 4.5.5 below. The diagonal terms of the turbulent fluctuations of the wall-normal and spanwise components ($i = j = 2$ and $i = j = 3$) were already discussed in Sec. 4.5.2 above in terms of turbulent fluxes, so that these are not addressed again here.

For scenario *HP*, the convective acceleration (term 2) is strongest in the near wall region and at the upper boundary (Fig. 4.19a). This is most pronounced at $z/H = 2$ as this is the location of the stable ridge. The one-dimensional profile shows high values in the bed-load layer

i	j	Term 2	Term 3	Term 4	Term 5	Term 6	Term 7	Term 8	Term 9
1	2	$7 \cdot 10^{-2}$	$1 \cdot 10^{-2}$	$1 \cdot 10^{-3}$	$1 \cdot 10^{-1}$	$7 \cdot 10^{-2}$	$1 \cdot 10^{-1}$	$1 \cdot 10^{-2}$	$1 \cdot 10^{-2}$
	3	$7 \cdot 10^{-2}$	0	0	$1 \cdot 10^{-2}$	$7 \cdot 10^{-2}$	$1 \cdot 10^{-3}$	$6 \cdot 10^{-3}$	$6 \cdot 10^{-3}$
2	2	$1 \cdot 10^{-3}$	0	0	$8 \cdot 10^{-2}$	$7 \cdot 10^{-3}$	$1 \cdot 10^{-4}$	$1 \cdot 10^{-4}$	$1 \cdot 10^{-4}$
	3	$1 \cdot 10^{-3}$	0	0	$4 \cdot 10^{-3}$	$1 \cdot 10^{-3}$	$1 \cdot 10^{-4}$	$8 \cdot 10^{-5}$	$8 \cdot 10^{-5}$
3	2	$1 \cdot 10^{-3}$	0	$4 \cdot 10^{-5}$	$1 \cdot 10^{-2}$	$1 \cdot 10^{-3}$	$5 \cdot 10^{-3}$	$3 \cdot 10^{-4}$	$3 \cdot 10^{-4}$
	3	$1 \cdot 10^{-3}$	0	0	$1 \cdot 10^{-2}$	$5 \cdot 10^{-4}$	$8 \cdot 10^{-3}$	$1 \cdot 10^{-4}$	$3 \cdot 10^{-4}$

Table 4.5: M *Maximum of the absolute intensity of the two-dimensional distribution of the locally averaged terms of the DAM momentum balance (4.34) for scenario HP.*

i	j	Term 2	Term 3	Term 4	Term 5	Term 6	Term 7	Term 8	Term 9
1	2	$5 \cdot 10^{-2}$	$7 \cdot 10^{-3}$	$5 \cdot 10^{-3}$	$7 \cdot 10^{-2}$	$5 \cdot 10^{-2}$	$3 \cdot 10^{-1}$	$6 \cdot 10^{-3}$	$6 \cdot 10^{-3}$
	3	$6 \cdot 10^{-2}$	0	0	$6 \cdot 10^{-3}$	$6 \cdot 10^{-2}$	$7 \cdot 10^{-4}$	$4 \cdot 10^{-3}$	$4 \cdot 10^{-3}$
2	2	$3 \cdot 10^{-3}$	0	0	$8 \cdot 10^{-2}$	$8 \cdot 10^{-3}$	$2 \cdot 10^{-4}$	$4 \cdot 10^{-5}$	$4 \cdot 10^{-5}$
	3	$8 \cdot 10^{-4}$	0	0	$4 \cdot 10^{-3}$	$7 \cdot 10^{-4}$	$2 \cdot 10^{-4}$	$2 \cdot 10^{-5}$	$2 \cdot 10^{-5}$
3	2	$1 \cdot 10^{-3}$	0	$6 \cdot 10^{-5}$	$3 \cdot 10^{-2}$	$1 \cdot 10^{-3}$	$6 \cdot 10^{-3}$	$7 \cdot 10^{-5}$	$7 \cdot 10^{-5}$
	3	$1 \cdot 10^{-3}$	0	0	$5 \cdot 10^{-3}$	$2 \cdot 10^{-4}$	$8 \cdot 10^{-3}$	$5 \cdot 10^{-5}$	$5 \cdot 10^{-5}$

Table 4.6:   *Maximum of the absolute intensity of the two-dimensional distribution of the locally averaged terms of the DAM momentum balance (4.34) for scenario LP.*

$0 < y < 0.2H$. In the outer flow, where the influence of particle-fluid interaction becomes negligible and a "clear-water" layer is present, the stresses reported in the two-dimensional plot average out to zero. This is also true for scenario *LP*, where the convective acceleration does not show such a distinct pattern.

Due to their linear dependency, the momentum supply (term 3) directly reflects the distribution obtained for ϕ_{VT} in Fig. 4.11a and b. With the total porosity being equal to unity in the outer flow field, the momentum supply is constant and decreases in the near-wall region as the total porosity decreases. The heterogeneity in the near-wall region is present for scenario *HP* (Fig. 4.20a), while homogeneous conditions are maintained throughout the channel for scenario *LP* (Fig. 4.20b). The regions with low momentum supply induce the areas of low speed fluid shown in Fig. 4.12a. Low porosity yields less momentum supply resulting in decelerated fluid.

As expected, the contributions from turbulent shear stresses are fairly constant and homogeneous in the outer flow for both scenarios (Fig. 4.21). In the near-wall region, the pattern of the ridges becomes obvious for scenario *HP*. Here, very low values prevail around $z/H = 1$ and $z/H = 3$ at $y = 0$, which is the area surrounding the stable ridge. The globally averaged profile shows two distinct peaks at $y = 0$ and $y/H = 0.1$, reflecting the geometry of the fixed bed and a first layer of particles either resting or rolling across the fixed bed. Again, this heterogeneity is not present for scenario *LP*.

The two-dimensional plot of the form-induced stresses depicted in Fig. 4.22 shows almost the same distribution as the convective acceleration in Fig. 4.19. This similarity arises from the correlated quantities involved in the local averaging procedure, if V_0 is comparatively small. Since term 2 and 6 both redistribute momentum supplied by term 3, they add up to

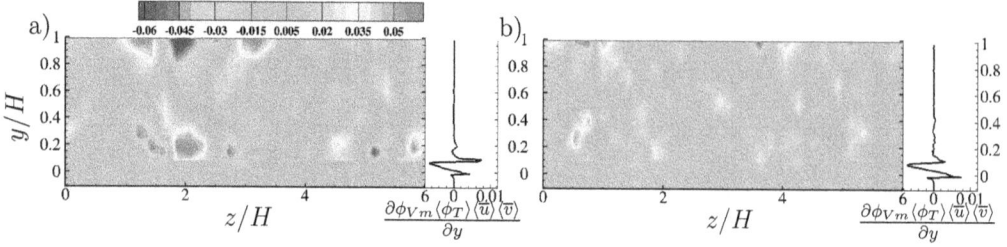

Figure 4.19:   *Spatial distribution of convective acceleration (term 2 of the DAM momentum balance. a) Scenario HP, b) scenario LP. Horizontal and vertical axis are not to scale.*

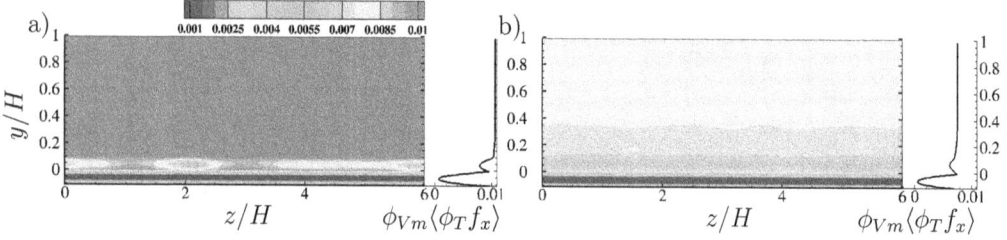

Figure 4.20:   *Spatial distribution of the momentum supply (term 3) for a) scenario HP and b) scenario LP. Horizontal and vertical axis are not to scale.*

the same effect instead of canceling each other out. Employing the large global averaging volume, the one-dimensional profiles of the two terms differ and the peaks around $y = 0.1H$ of term 6 are less pronounced than for term 2.

The local distribution of term 7 shows two elevations, where the viscous stress is sizable for scenario *HP* (Fig. 4.23a): one at the interface of the fixed bed and a second one at $y = 0.11H = 1D$ representing the stresses induced by resting particles. This term is the one with the highest magnitudes among the quantities investigated, which is in agreement with studies of turbulent channel flows across smooth walls [93, 85, 73]. The heterogeneity in spanwise direction once more becomes evident for scenario *HP*, while scenario *LP* shows homogeneous conditions in spanwise direction as all the particles are in motion.

4.5.4   Momentum balance

The momentum balance (4.34) can be determined with yet another averaging operator, involving blockwise averaging in vertical direction. This is done now to provide information going beyond Tab. 4.5 and 4.6, even in the presence of mobile sediment. To this end, an averaging volume over the streamwise and spanwise extent of the computational domain is considered, starting at some level y and reaching to the upper boundary ($y = H$). The free-slip boundary condition imposed at the top of the domain reads $\partial u / \partial y = \partial w / \partial y = 0$ and $v = 0$. Since furthermore derivatives in x- and z-direction of globally averaged quantities vanish, only terms with index $i = 1$ and $j = 2$ are different from zero. Integration of (4.34) with $i = 1$ over the wall-normal direction from some distance y to the upper boundary, hence

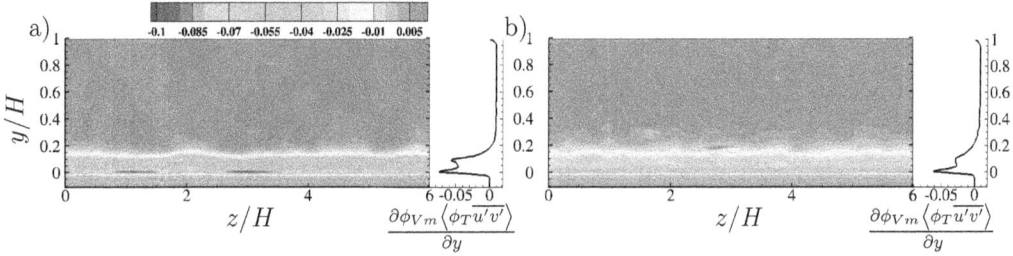

Figure 4.21: *Spatial distribution of turbulent stresses (term 5) for a) scenario* HP *and b) scenario* LP. *Horizontal and vertical axis are not to scale.*

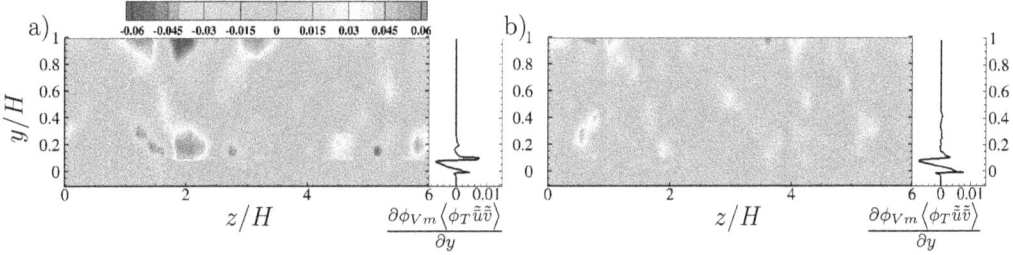

Figure 4.22: *Spatial distribution of form-induced stresses (term 6). Left: Scenario* Heavy, *right: Scenario* Light. *Horizontal and vertical axis are not to scale.*

yields

$$-\underbrace{\phi_{Vm}\langle\phi_T\rangle\langle\overline{u}\rangle\langle\overline{v}\rangle}_{2} = \overline{f}_x\underbrace{\int_y^H \phi_{Vm}\langle\phi_T\rangle dy^*}_{3} + \underbrace{\phi_{Vm}\left\langle\phi_T\overline{u'v'}\right\rangle}_{5} + \underbrace{\phi_{Vm}\left\langle\phi_T\tilde{\overline{u}}\tilde{\overline{v}}\right\rangle}_{6}$$

$$-\underbrace{\phi_{Vm}\left\langle\phi_T\nu\overline{\frac{\partial u}{\partial y}}\right\rangle}_{7} + \underbrace{\phi_{Vm}\langle\phi_T\tilde{\overline{u}}\rangle\langle\overline{v}\rangle}_{8} + \underbrace{\phi_{Vm}\langle\phi_T\tilde{\overline{v}}\rangle\langle\overline{u}\rangle}_{9}$$

$$+\underbrace{\int_y^H \frac{1}{\rho}\frac{1}{V_0}\overline{\iint_{S_{int}} pn_x dS}^s dy^*}_{10} - \underbrace{\int_y^H \frac{1}{V_0}\overline{\iint_{S_{int}}\left(\nu\frac{\partial u}{\partial y^*}\right)n_y dS}^s dy^*}_{11} - \underbrace{\int_y^H \frac{\partial\phi_{Vm}\langle\phi_T\overline{u}\rangle}{\partial t}dy^*}_{1} \quad .$$

$$(4.40)$$

with $n_x = n_1$ and $n_y = n_2$, as defined in (4.32). The spatially averaged quantities $\langle...\rangle$ are functions of the wall-normal coordinate y, which is not explicitly indicated here for clarity. Division by the averaging volume is omitted as well. In [115], it was argued that only the terms 3, 5, 6, 10, and 11 are relevant for fixed bed conditions with high submergences, while the terms 2, 7, 8, 9 can be neglected. For the case of a mobile granular bed, this assumption does not necessarily hold, since particles are free to move and may cause momentum fluxes in the wall-normal direction of the domain. Hence, all these terms are accounted for in the present analysis in order to fully explore the importance of the different terms of Eq. (4.34). As discussed in Sec. 4.4.3 and 4.5.3, the remaining terms of (4.40) were not computed directly. Since the total momentum balance must add up to zero, the sum of the evaluated terms must be the "out-of-balance" term comprising the local acceleration (term 1) as well

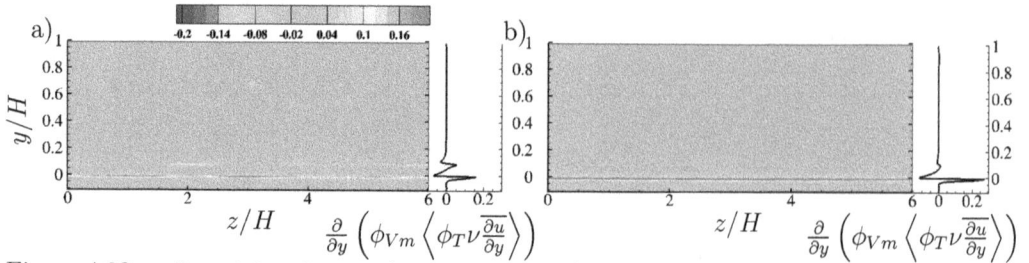

Figure 4.23:   *Spatial distribution of viscous stresses (term 7) for a) scenario HP and b) scenario LP. Horizontal and vertical axis are not to scale.*

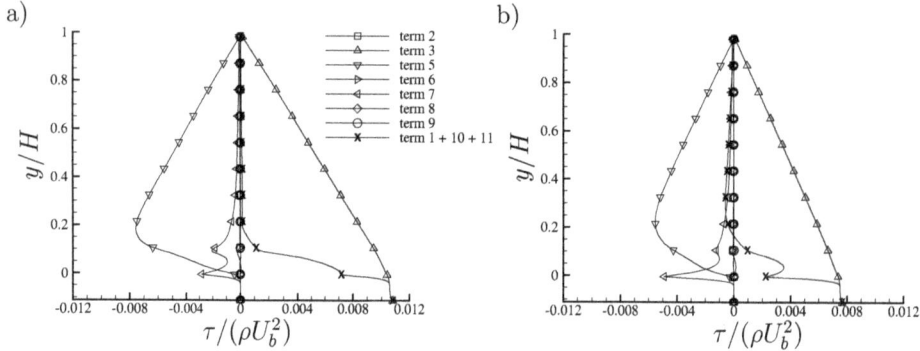

Figure 4.24:   *Wall-normal distribution of the different terms of the integral form of the momentum balance. a) Scenario HP, b) Scenario LP. Vertical distances between symbols correspond to D.*

as the interfacial terms (term 10 and 11).

Fig. 4.24 displays the importance of the different terms of the total momentum balance summarizing the computations presented in Sec. 4.5.2 and 4.5.3. The most dominant contribution to the total stress originates from the turbulent fluctuations (term 5) balancing the driving external force, i.e. term 3. As it was discussed by [140] and [157, 155], turbulent stresses are amplified by the introduction of inertial particles, which are much larger than the viscous length scales. The magnitude of the viscous stress (term 7), on the other hand, is diminished compared to the case of a turbulent channel flow with smooth walls as presented in [93]. Although, it is highlighted in Sec. 4.5.3 that form-induced stresses (term 6) can contribute to the total stress budget in the near-wall region, their contribution remains small if global averaging is performed. As discussed in Sec. 4.5.3 above, the contribution becomes more significant for a local averaging volume resolving heterogeneities in spanwise direction. The remaining terms were found to have little or no impact on the momentum balance, neither for global nor for local averaging. Assuming term 1 to be close to zero, the out-of-balance term is equal to the surface terms 10 and 11 and represents the stress taken up by the particles in the near-wall region. This quantity has two distinct kinks at $y = 0$ and $y = 0.11H = 1D$. As expected, the mobile particles of scenario HP, located mainly between $0 < y < 0.1H$, take up the gross part of the momentum supplied by term 3, whereas for scenario LP the strongest gradient of the surface terms is reported for the interval $-0.05H < y < 0$, i.e. within the interstitial of the fixed bed. In the outer flow with no particles being present, the surface term becomes zero for scenario HP. This proves that sufficient data was gathered to reach statistical convergence and the local acceleration actually is zero as postulated in Sec. 4.4.3. For scenario LP, a small closure gap remains in the outer flow that can only be due to local acceleration, since no particles are present in this area and term 10 and term 11 vanish.

Choosing $y = -D$ in (4.40), i.e. integrating over the full wall-normal extent of the channel, terms 1, 2, 5, 6, 8, and 9 of (4.40) vanish due to the boundary conditions at the lower wall of the domain. Term 1 is neglected due to averaging over $T_0 = T_{sample}$. This simplifies Eq. (4.40) to

$$\overline{f}_x \int_{-D}^{H} \phi_{Vm} \langle \phi_T \rangle \mathrm{d}y^* = \int_{-D}^{H} \overline{\frac{1}{V_0} \iint_{S_{int}} \left(\nu \frac{\partial u}{\partial y^*} \right) n_y \mathrm{d}S}^{s} \mathrm{d}y^* - \int_{-D}^{H} \overline{\frac{1}{\rho} \frac{1}{V_0} \iint_{S_{int}} p n_x \mathrm{d}S}^{s} \mathrm{d}y^*$$
$$- \phi_{Vm} \left\langle \phi_T \nu \overline{\frac{\partial u}{\partial y}} \right\rangle \bigg|_{y^*=-D} \quad . \tag{4.41}$$

The last term on the right-hand side represents the integrated friction at the bottom of the domain, but is very small since the vertical gradient of $\langle \overline{u} \rangle$ is close to zero at $y = -D$, as demonstrated by the result displayed in Fig. 4.12 above. Conceptually, this is also an interfacial term as it results from the momentum exchange of the fluid at the interface with the solid bottom wall. The terms on the right-hand side of (4.41) determine the resistance of the channel, hence the total stress, and are balanced by the momentum supply of the channel. As mentioned above, computation of the right hand side of (4.41) is not possible with the present data, but due to the balance in (4.41) the evaluation of the left-hand side of this equation is possible and can be used to define a wall shear stress $\tau_w^{(p)}$ by setting

$$\frac{\tau_w^{(p)}(\Lambda)}{\rho_f} = \overline{f}_x \int_{-D}^{H} \phi_{Vm} \langle \phi_T \rangle \mathrm{d}y^* \quad . \tag{4.42}$$

This quantity represents the bed-shear stress projected onto a horizontal plane $\Lambda = [0; L_x[\times [0; L_z[$, highlighted by the argument (Λ) to distinguish it from locally averaged values. It is induced by the bottom wall, the fixed particles, and the moving granular bed altogether [116] and yields the total stress taken up by the phase boundary in a straightforward manner. From this analysis, an *a posteriori* friction velocity $u_\tau^{(p)} = \sqrt{\tau_w^{(p)}/\rho_f}$ of the particle-laden flow can be determined and used to define the relevant dimensionless *a posteriori* quantities of the flow with bed-load transport , which are collected in Tab. 4.7. This gives an integral measure of the additional hydraulic resistance introduced by the mobile bed. While the particle Reynolds number D^+ is 19.2 for the unladen rough-bed scenario, it is increased to $D^{+(p)} = 31.2$ in scenario HP. Hence, increasing the amount of mobile particles by 30% of one layer of fixed spheres yields an increase of the bed shear stress by 62%. This proves that the patterns of inactive particles described in Sec. 4.2.2 act as roughness elements substantially enhancing the hydraulic resistance of the channel. For scenario LP with the same number of mobile particles, just lighter, the friction Reynolds number is $D^{+(p)} = 26.2$, which is 36% larger than the value for the unladen flow. Hence, the increase in shear stress is almost equivalent to the increase of the interface area S_{int} between the particles and the fluid. It is important, to highlight, that this *a posteriori* value of $D^{+(p)}$ does not give a suitable measure for the required resolution of the DNS. Instead, it was reported by [149] that a discretization of $D/\Delta_x = 20$ is sufficient to simulate particle Reynolds numbers with values of up to 136 based on the slip velocity, i.e. the absolute value of the particle velocity with respect to the mean fluid velocity. This is well above the present regime as demonstrated in [151], so that

Scenario	$\tau_w^{(p)}/(\rho_f U_b^2)$	$u_\tau^{(p)}/U_b$	$D^{+(p)}$	$Re_\tau^{(p)}$
HP	$1.08 \cdot 10^{-2}$	0.104	31.2	281
LP	$0.76 \cdot 10^{-2}$	0.087	26.2	236

Table 4.7:   A posteriori *dimensionless numbers for the present mobile bed conditions.*

even when based on $D^{+(p)}$, the spatial resolution by the DNS grid is sufficiently fine. Finally, note that computing a modified *a posteriori* Shields number is not appropriate, as the concept of [141] applies to the threshold of mobilization, and not the situation of a sizable amount of sediment being transported. It furthermore supposes uniform roughness which is no longer the case with non-uniform distribution of particles accumulating in clusters, such as the ridges encountered with scenario *HP*.

4.5.5 Secondary currents

It is known that turbulent flows over streamwise ridges of resting particles, as observed in scenario *HP*, typically go along with the creation of secondary currents in form of counter-rotating cells, a feature referred to as Prandtl's second kind of secondary flow, caused by anisotropy in turbulence [21]. This introduces a mean streamwise vorticity

$$\langle \overline{\omega}_x \rangle = \frac{\partial \langle \overline{w} \rangle}{\partial y} - \frac{\partial \langle \overline{v} \rangle}{\partial z} \tag{4.43}$$

with absolute values of the wall-normal and spanwise velocity component that have a magnitude of up to 5% of the bulk velocity [110]. Each of the cells of the secondary flow has an extent of H in wall-normal and spanwise direction resulting in the typical spacing of $2H$ of ridges. Indeed, these characteristics can be observed for scenario *HP* for term 3 in Fig. 4.20 and term 5 in Fig. 4.21.

The impact of secondary flows on the present scenarios is now addressed employing the DAM framework with local averaging in space, i.e. using $V_{0,2}$ according to (4.38). Similar to Equation (4.40), integrating (4.34) over the wall-normal direction for $i = 1$, the double-averaged momentum balance gives the relevant terms causing the lateral momentum exchange. Based on the analysis of the different terms in Sec. 4.5.3, the momentum flux due to spatial correlation (term 8 and 9) were not included. Moreover, contributions of the interfacial terms (term 10 and 11) were neglected, because the secondary flow predominantly evolves in the free fluid, where no particles are present as proposed in [41, 115]. These simplifications allow for the following approximation

$$
\overline{f}_x \int_y^H \phi_{Vm} \langle \phi_T \rangle \mathrm{d}y^* \approx \int_y^H \left[\frac{\partial \phi_{Vm} \langle \phi_T \rangle \langle \overline{u} \rangle \langle \overline{v} \rangle}{\partial y^*} + \frac{\partial \phi_{Vm} \langle \phi_T \rangle \langle \overline{u} \rangle \langle \overline{w} \rangle}{\partial z} \right] \mathrm{d}y^* +
$$

$$
\int_y^H \left[\frac{\partial \phi_{Vm} \langle \phi_T \overline{u'v'} \rangle}{\partial y^*} + \frac{\partial \phi_{Vm} \langle \phi_T \overline{u'w'} \rangle}{\partial z} \right] \mathrm{d}y^* + \quad (4.44)
$$

$$
\int_y^H \left[\frac{\partial \phi_{Vm} \langle \phi_T \tilde{\overline{u}}\tilde{\overline{v}} \rangle}{\partial y^*} + \frac{\partial \phi_{Vm} \langle \phi_T \tilde{\overline{u}}\tilde{\overline{w}} \rangle}{\partial z} \right] \mathrm{d}y^* \quad .
$$

The difference with respect to (4.40) above is that in (4.44) local averaging in space is used, while global averaging is employed in (4.40). A flow without secondary currents fulfills the condition $\overline{v} = \overline{w} = 0$ driving the streamwise vorticity (4.43) to zero. Assuming furthermore statistical homogeneity in z reduces (4.44) to

$$
\overline{f}_x \int_y^H \phi_{Vm} \langle \phi_T \rangle \mathrm{d}y^* = \int_y^H \left[\frac{\partial \phi_{Vm} \langle \phi_T \overline{u'v'} \rangle}{\partial y^*} + \frac{\partial \phi_{Vm} \langle \phi_T \tilde{\overline{u}}\tilde{\overline{v}} \rangle}{\partial y^*} \right] \mathrm{d}y^* \quad . \quad (4.45)
$$

The remaining part of the shear stress, hence, can be attributed to stresses induced by secondary currents. The DAM framework now allows evaluating this contribution in a consistent manner and suggests to define

$$
\frac{\tau_{sec}}{\rho_f} = \int_y^H \left[\frac{\partial \phi_{Vm} \langle \phi_T \rangle \langle \overline{u} \rangle \langle \overline{v} \rangle}{\partial y^*} + \frac{\partial \phi_{Vm} \langle \phi_T \rangle \langle \overline{u} \rangle \langle \overline{w} \rangle}{\partial z} \right] \mathrm{d}y^* +
$$

$$
\int_y^H \left[\frac{\partial \phi_{Vm} \langle \phi_T \overline{u'w'} \rangle}{\partial z} + \frac{\partial \phi_{Vm} \langle \phi_T \tilde{\overline{u}}\tilde{\overline{w}} \rangle}{\partial z} \right] \mathrm{d}y^* \quad (4.46)
$$

as an integral part of the total stress $\tau_w^{(p)}$. With the local average applied here and y being an independent variable in (4.46), τ_{sec} is a function of z and y and can be plotted with respect to these coordinates, similar to the figures before, as shown in Figure 4.25. These graphs also provide vector plots of the mean secondary flow, using the global averages as a reference defining

$$
v_{sec} = \langle \overline{v} \rangle_{V_{0,2}} - \langle \overline{v} \rangle_{V_{0,1}} \quad (4.47a)
$$

$$
w_{sec} = \langle \overline{w} \rangle_{V_{0,2}} - \langle \overline{w} \rangle_{V_{0,1}} \quad , \quad (4.47b)
$$

with the local averaging volume denoted by the subscript $V_{0,2}$ and the global averaging volume by $V_{0,1}$, respectively.

For scenario HP, the vector plot in Figure 4.25a shows three cells that extend over the entire channel height and indicate secondary currents of different intensity. The most pronounced one is located at $z = 2H \pm 1H$, which is also the position of the stable ridge. Another cell is located at $z = 0H \pm 1H$ reaching over the periodic boundary from $z = 5H$ to $z = 1H$. The cell is weaker because it was shown that the roughness element is constituted of a finite large-scale cluster that moves slowly in streamwise direction. The third cell around $z = 4H \pm 1H$

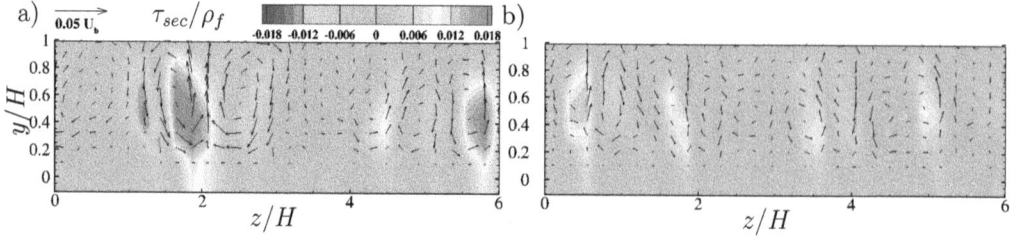

Figure 4.25:   *Secondary currents induced by light particles represented by the vector $(w_{sec}, v_{sec})^T$. Spatial distribution of τ_{sec} and wall normal plot of the integral form of τ_{sec} for a) scenario* HP *and b) scenario* LP. *The reference vector provides a reference to a magnitude of 0.05 U_b. Horizontal and vertical axis are not to scale.*

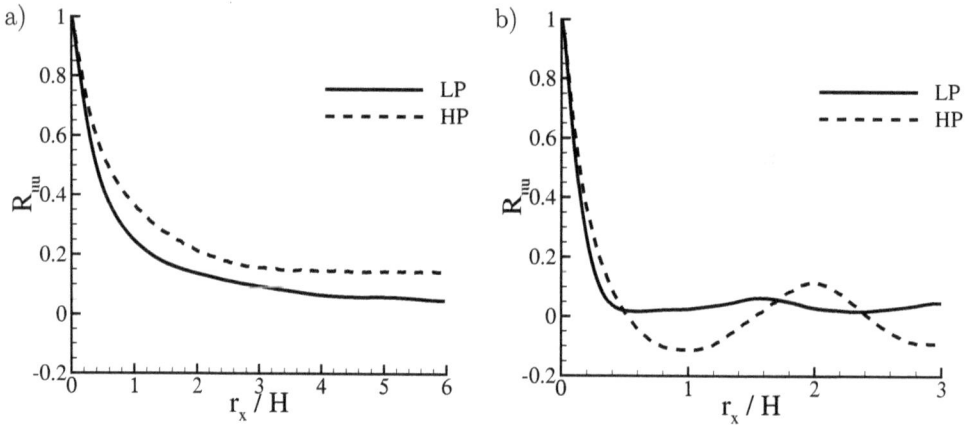

Figure 4.26:   *Two-point correlation function of the fluid phase at $y = 0.5D$.   a) Streamwise direction, b) Spanwise direction.*

is very disturbed due to the irregular particle clustering. In the free flow above the sediment, large positive values of τ_{sec} correspond to strong upward fluid motion, while negative values indicate downward moving fluid. The light particles of scenario LP induce weaker secondary currents (Figure 4.25b). The cells in this case have a smaller spanwise extent ($\approx 1.5H$) and do not reach over the entire channel height.

For the simulations presented, this illustrates that not only do the particles accumulate in low-speed streaks, as it has been suggested by other experimental and numerical studies with particles of small size, i.e. point particles, [100] and particles of finite size [91], but the particle-fluid interaction completely re-organizes the flow reflecting the state of equilibrium for the given physical parameters. For the present scenarios, stationary cells of secondary currents develop with the introduction of heavy particles. These cells are bounded by particle clusters that move in streamwise direction with time scales much larger than the turbulent fluid time scale. Particles with a mobility higher than the threshold proposed by [141] disturb this pattern and shift the lower bound of the cells of secondary flow upwards.

The modificaton of the large-scale coherent fluid structures was also confirmed in [157] by means of the globally averaged two-point-correlation analysis

$$R_{uu}(r_x) = \frac{\left\langle \overline{u'(x)u'(x+r_x)} \right\rangle}{\sqrt{\left\langle \overline{u'(x)u'(x)} \right\rangle \left\langle \overline{u'(x+r_x)u'(x+r_x)} \right\rangle}} \tag{4.48}$$

in streamwise direction and spanwise direction using r_x and r_z as the respective component of the distance vector $\mathbf{r}$. Just as the particles intensify the fluctuations, they increase the size of the coherent structures for particle laden flow. In streamwise direction, the extension of the coherent structures leads to a development of the two-point-correlation function for the layer $y = 0.5D$ that does not decay to zero (Fig. 4.26a). This is true for both scenarios LP and HP. For the latter, this effect is due to the stability of the ridges, that form channels in between with a distinct secondary current. For scenario LP, however, this was not expected a priori, because of the random distribution. Nevertheless, the light particles create an increasing coherence of the flow field, which supports the observations stated above. A strong secondary peak of the two-point-correlation function at $r_z \approx 2H$ was reported in the near-wall region for scenario HP in Fig. 4.26b, which can clearly attributed to the secondary currents discussed above and illustrates once more the impact of the particles on the overall flow behavior.

4.6   Conclusions

Inertial, cohesionless particles of finite size forming a granular bed create a variety of multi-scale spatial heterogeneities of the bed morphology that may significantly affect the flow structure and particle-fluid interaction. Thus, interrelations between flow and mobile particles have to be studied within a framework coupling flow and particles in a rigorous way, which is also flexible enough to account for the individually transported particles as well as clusters of various kinds. In this paper, for the first time the framework of the Double-Averaging-Methodology was employed to provide a full description of a flow with mobile, granular beds. The most prominent quantities obtained with Direct Numerical Simulations, such as turbulent stresses and form-induced stresses, were calculated and analyzed employing two different averaging strategies distinguished by the size of the spatial averaging domain. Two simulation scenarios were addressed: the first features heavy particles with their mobility below the conventional threshold of motion. The second scenario involves light particles with mobility well above the threshold of motion.

A full description of the momentum balance could be given by computing all the individual stress terms related to the fluid. It was shown that form-induced stresses can contribute a significant part to the momentum balance and can locally become quite strong. This underlines the necessity of an averaging volume that is able to capture the spatial hetero-geneities in the flow introduced by mobile and fixed particle clusters usually addressed as morphological forms. Using different sampling strategies, it was possible to highlight differ-ent key flow characteristics such as mobile bed layers, stable particle clusters, flow resistance to increased frictional forces, and secondary currents. Examples are given of how to use the double-averaged equations to describe these flow features. The observed patterns are more pronounced for the case of heavy particles, where a separation of the morphological scales from the turbulent time scales takes place, while they are less pronounced for the case of light particles. The analysis highlights the capability of the DAM to describe particle-laden flows and introduces a variety of statistical quantities which can now be applied to other simulation raw data of this type. The generated results may serve as a knowledge base for improving existing flow models for mobile-bed conditions.

5 Impact of mobility and sediment supply on bed-load transport

5.1 Introduction

Predicting sediment transport is an important task for environmental engineers as the river geometry is determined by the net rate of deposition and erosion on the sediment bed. Scientists have tackled the issue of sediment transport for more than a century [24]. Shields in 1936 was among the first, who formulated a framework for thresholds of incipient motion [141]. In this study, a dimensionless parameter was introduced comparing the average frictional force to the gravitational forces inhibiting the mobilization of a particle embedded in a sediment packing today called the Shields number (3.13b). On the basis of his work, Meyer-Peter and Müller (1948) derived formulas for predicting sediment transport at high Reynolds numbers and high submergence. This approach, however, neglects particle-particle interaction within large scale mobile clusters as well as the heterogeneity introduced by non-homogeneous particle distributions.

The resulting bed forms reflecting the equilibrium between erosion and deposition are diverse, however. At field scale, streamwise-oriented ridges were reported by Karcz [84] for small flow velocities and small mass loading. These structures are induced by secondary currents forming vortex tubes in and over the troughs [9]. The sediment pattern can be modified substantially by varying the sediment supply. Indeed, different types of particle structures can be witnessed in experimental flumes for similar uniform flow conditions, when the mass loading is increased [47]. In the study presented in this reference, the relative density of the sediment was close to the critical threshold of incipient motion based on the condition of Shields [141]. Runs with low sediment supply produced ridges, confirming the observations reported above. For larger mass loadings on the other hand, spanwise oriented dune-like structures were observed. Due to the multi-disperse sediment used for these experiments, preferential transport changed the constitution of the bed load for different runs and exact distinction of the different physical mechanisms was not possible [127].

The complex physics of sediment transport raise the need for highly resolved data under controlled flow conditions [108]. Only recently sufficient computational resources have become available to conduct Direct Numerical Simulations (DNS) of such phenomena, which offer access to the desired detailed data. As discussed in the introduction, bed-load transport is characterized by locally very high mass fractions and dominance of interaction by collisions and contact. Hence, a sophisticated four-way coupling of the flow is needed to simulate such dense systems properly [17]. This requirement is met by the Immersed Boundary Method (IBM), which has proven to be a valuable tool for conducting efficient simulations of multi-phase flows that require full resolution and full coupling of the disperse phase and the fluid phase [83, 148, 88].

Computational studies of this kind on a large scale are still rare as the resolution requirements lead to very costly simulations of large domains over long time intervals. As a matter
of fact, only a few years ago it was doubted that this was feasible at all at the required scale
[17]. Hence, the goal of the chapter is to take the state of the art one step further. The scenario considered here is the particle-laden flow over a rough wall using the Adaptive Collision
Model (ACM) [87]. It was shown in Chap. 3 that this model has the capability to represent
sediment close to the mobilization threshold allowing for simulations of particle-laden flows
with contact-dominated bed-load transport. The size of the computational domain is, to the
knowledge of the author, the largest employed so far for this problem. The large domain allows the disperse phase to develop bed forms with high fidelity, hence giving detailed insight
into mechanisms of particle–fluid interaction. The characteristic numbers of the disperse
phase considered here are similar to those of experiments in laboratory at medium Reynolds
number [47, 142, 29]. The goal of the study is to reproduce patterns that are known from
experimental evidence at a Reynolds number of the transitionally rough regime. Using the
DAM-framework, the modification of turbulent fluctuations as well as the hydraulic resistance is assessed. Afterwards, this modification is explicitly linked to the time-averaged
particle patterns and bed elevations investigated by suitable statistical tools.

Most of the results of the following sections were first presented at the *8th International Con-
ference on Multiphase Flow* [154] and the *9th ERCOFTAC Workshop Direct and Large Eddy
Simulations* [158]. Moreover, the work was selected to be part of a special issue of *Advances
in Water Resources* [155] and for the Symposium of the *Neumann-Institut for Computing* at
the *Jülicher Supercomputing Center* [156].

5.2   Numerical setup

5.2.1   Computational domain and key parameters

The considered setup consists of an open-channel flow as outlined in Chap. 2 that features
periodic boundary conditions in streamwise and spanwise direction together with a free-slip
condition at the top and a no-slip condition at the bottom of the domain and on the particle
surface as illustrated in Fig. 5.1. The mean flow is driven by a streamwise pressure gradient
implemented as a volume force $\mathbf{f} = (f_x(t), 0, 0)^T$ in Eq. (2.1) as defined in Sec. 2.2.5. Hence,
the bulk velocity and the fluid mass flux are kept constant and the hydraulic resistance of
the channel in terms of the friction velocity u_τ can vary between different configurations.

A rough wall at the bottom is modeled by a layer of hexagonally packed, fixed mono-disperse
spheres with diameter D lying on the bottom boundary. For comparison, a case with an
additional layer of immobile particles is considered (case *Fix*), so that the sediment bed has a
thickness of $H_{sed} = 1.8D$ in this case as shown in Fig. 5.2. The extent of the computational
domain is $L_x \times L_y \times L_z = 24H \times (H + H_{sed}) \times 6H$ with $H = 9D$ being the water depth
and $H_{sed} = 1.8D$ in all cases. The streamwise extent L_x and the spanwise extent L_z were
chosen as multiples of the typical wavelength of the coherent structures observed for turbulent flows over rough walls being $\lambda_x = 6H$ and $\lambda_z = 2H$, respectively, [160] to guarantee
the development of large scale flow structures as well as particle clusters with high fidelity.
The origin of the vertical coordinate was set to the top of the upper layer of the fixed bed
and maintained for all cases with mobile particles at this distance from the lower domain
boundary (Fig. 5.2). The friction velocity u_τ was calculated by extrapolating the total shear

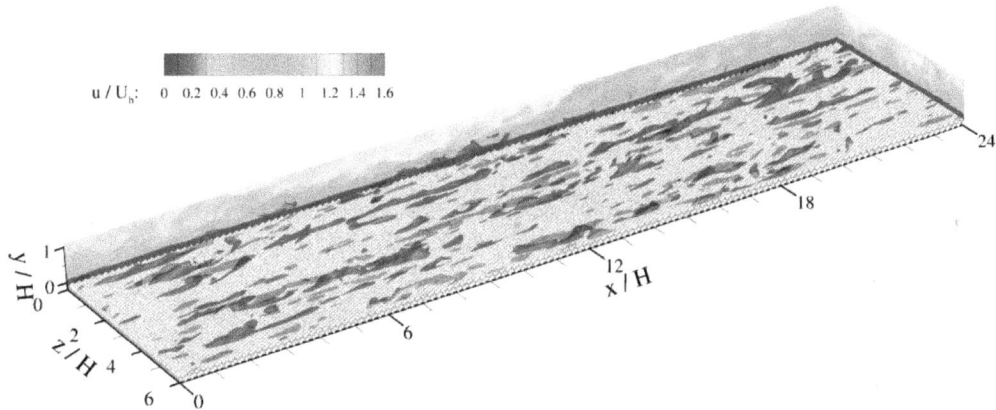

Figure 5.1: *Computational domain with fixed sediment bed (case* Fix*) and instantaneous values at an arbitrary instant. Contour plot of u/U_b on the sides of the domain, 3d-iso-surfaces of fluid fluctuations with $u'/U_b = -0.3$ in blue and $u'/U_b = 0.3$ in red inside the domain.*

Scenario	H/D	Re_b	$\tau_w/(\rho_f U_b^2)$	u_τ/U_b	Re_τ	D^+	Fr
present	9	2941	$0.513 \cdot 10^{-2}$	0.065	193	21.1	0.35
Cameron *et al.* [29]	8.8	3114	$0.440 \cdot 10^{-2}$	0.066	201	22.8	0.48

Table 5.1: *Comparison of the present numerical setup with the experiments of [29].*

stress profile of case *Fix* down to $y = 0$. Hence, u_τ is a measure of the shear rate at the interface between the fixed particles and the flow. With a mobile sediment bed, the definition of a shear velocity becomes delicate since the particles transfer additional momentum to the fluid by collisions with the rough wall. For this reason, the value obtained from the immobile sediment (case *Fix*) is also used in the cases with mobile sediment to determine the Shields number, normalization, and perform comparison with data from literature.

The physical parameters of the present simulations are summarized in Tab. 5.1. The bulk Reynolds number $Re_b = U_b H/\nu_f$ is 2941 with the bulk velocity defined as $U_b = \frac{1}{H} \int_0^H \langle u \rangle (y) \mathrm{d}y$. Angular brackets indicate averaging as specified below. The simulation with fixed bed yields a Reynolds number based on the friction velocity of $Re_\tau = u_\tau H/\nu_f = 193$ and a particle Reynolds number $D^+ = u_\tau D/\nu_f = 21$, which falls into the transitionally rough regime [77]. The simulations were conducted with an equidistant Cartesian grid of $4800 \times 240 \times 1198$ cells. This yields a total number of approximately $1.4 \cdot 10^9$ grid cells, and a discretization of each particle with $D/\Delta_x = 22$ points per diameter. The resolution in terms of wall units is $\Delta_x^+ = u_\tau \Delta_x/\nu = 0.95$, which is sufficient to resolve the smallest scales in the near-wall region as illustrated by the close-up shown in Fig. 5.3.

In addition to case *Fix* described above, five runs with mobile particles were performed using the same Cartesian grid to investigate different flow conditions. The common parameter values are assembled in Tab. 5.2, the specific ones in Tab. 5.3. Mobile sediment particles were initialized by positioning them at some elevation above the rough wall. For the first run serving as a reference (case *Ref*), the number of mobile particles constituting the mass loading was $N_{p,mob} = 13500$, which is equivalent to one full layer of the hexagonally packed spheres fixed at the bottom. In the following, these mobile particles will be termed "active"

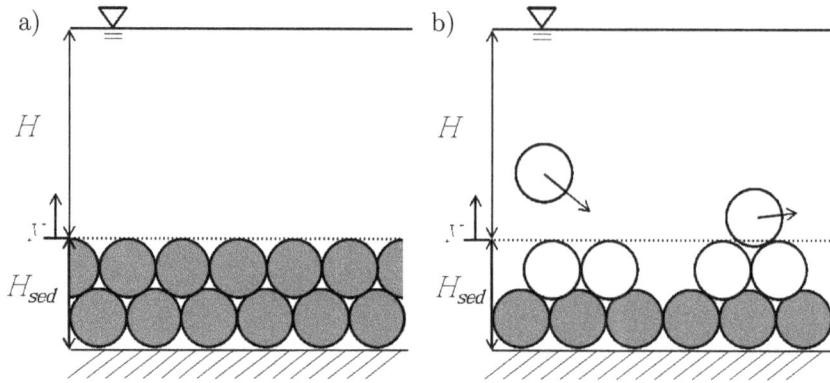

Figure 5.2: *Sketch of the computational setup with immobile particles (gray) and mobile particles (white). a) Fixed sediment bed (case* Fix*) b) cases with rough wall made of immobile particles and mobile particles (*Ref, FewPart, ManyPart, LowSh, *and* HighSh*).*

L_x/H	L_y/H	L_z/H	D/Δ_x	Δ_x^+	N_x	N_y	N_z	N_{tot}
24	1.202	6	22.2	0.95	4800	240	1198	$1.4 \cdot 10^9$

Table 5.2: *Common numerical parameters of all simulations.*

if they are moving with a velocity larger than the friction velocity u_τ defined above. Particles temporally resting on the bottom wall are termed "inactive". The restitution coefficient of the mobile particles describes the amount of damping during collisions and is defined by $e = -u_{p,out}/u_{p,in}$, with $u_{p,in}$ the particle velocity before and $u_{p,out}$ the velocity after the collision [87]. Its value was set to $e = 0.97$ corresponding to the one of glass beads [80], which hence is close to the one of sand grains.

To quantify the mobility of the particles, the Shields number $Sh = u_\tau^2/(\rho'gD)$ is used, with $\rho' = (\rho_p - \rho_f)/\rho_f$ being the relative particle density. For the reference run, the relative density was chosen such that the Shields parameter is $Sh = 0.04$. This value is slightly above the critical value of incipient motion, $Sh_{crit} = 0.034$, which was extracted from the empirical graph of Shields in [141]. In their review [26], Buffington and Montgomery stress that the sediment supply is an important parameter in the present type of configuration. To address its impact, the simulation with the parameters of case *Ref* was repeated by inserting only half the number of mobile particles, i.e. $N_{p,mob} = 6750$ on the one hand (case *FewPart*) and by inserting twice as many particles, i.e. $N_{p,mob} = 27000$ on the other (case *ManyPart*). *Ceteris paribus*, a comparison between the runs gives a valuable information about the formation of particle structures as a function of mass loading, adding to the findings of Dietrich *et al.* [47].

Finally, the relative density ρ' was increased by 55% in order to obtain a Shields number $Sh/Sh_{crit} = 0.75$ nominally below the critical value of sediment motion for case *LowSh*. This allows to investigate turbulence modifications as a function of mobility of the sediment [140] and the role of extreme flow events on particle mobilization as discussed in [32]. Moreover, a complementary run with a Shields number $Sh/Sh_{crit} = 1.82$ well above the threshold was carried out to address the collective motion of a granular bed that is fully active (case *HighSh*). All five simulations were initialized in the same way and run until the erosion and deposition rates were in equilibrium. This was verified using an estimation procedure

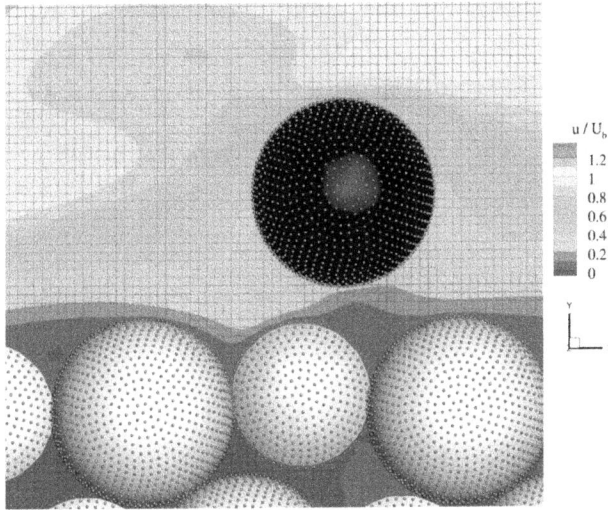

Figure 5.3: *Illustration of the IBM discretization as applied in the present simulations. Particle surfaces with discrete marker points. Straight lines represent the Cartesian background grid. The contour plot shows an instantaneous snapshot of u/U_b and was taken from the case LowSh.*

Case	$N_{p,fix}$	$N_{p,mob}$	ρ'	Sh/Sh_{crit}
Fix	27000	0	–	–
Ref	13500	13500	0.116	1.18
FewPart	13500	6750	0.116	1.18
ManyPart	13500	27000	0.116	1.18
LowSh	13500	13500	0.182	0.75
HighSh	13500	13500	0.074	1.82

Table 5.3: Specific parameters of the simulations reported.

described in [154] (cf. App. C). In the initial phase, inserting the particles at a level with full exposure to the turbulent flow creates a disturbance of the driving volume force f_x. This disturbance is dampened out after some time t_{init} as illustrated by the time series of f_x in Fig. 5.4. The averaging procedures described below were applied to the sampling interval with steady-state conditions, i.e. for $t - t_{init} > 0$, and conducted over more than 260 bulk units $T_b = H/U_b$. Note that $f_x(t)$ in Fig. 5.4 was filtered by a moving average window of $6H/U_b$. This time scale corresponds to the large scale flow structures reported in [160]. The simulations were carried out on an IBM BlueGene/Q consuming more than 70 million CPU-h.

5.2.2 Comparability to experimental setups

Due to the computational requirements of a "true" DNS, the present study is still limited to particles with a low Reynolds number and a low submergence H/D. It is, therefore, comparable to experimental studies that used oil [28, 29] or sugar solution [40] as a fluid. The present setup is based on the flume design of [29], where oil with a kinematic viscosity of $\nu_f = 15 \cdot 10^{-6} m^2/s$ and immobile glass spheres of diameter $D = 11.1mm$ in a hexagonal

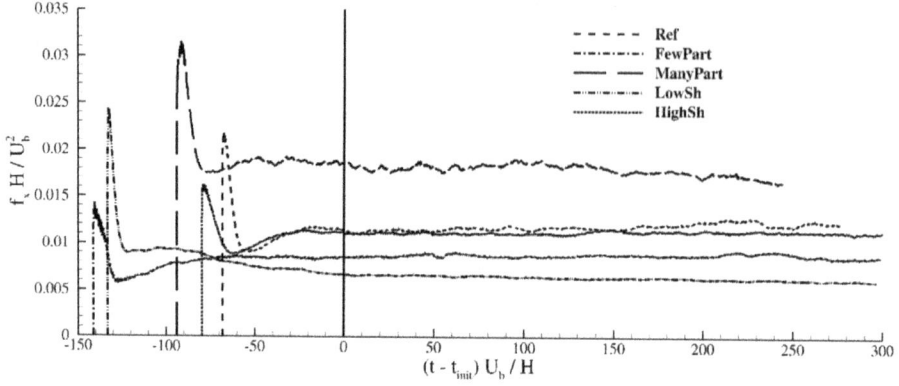

Figure 5.4: *Time evolution of the volume force driving the channel flow filtered with a moving average window of $6H/U_b$. The vertical line indicates the beginning of the sampling period.*

arrangement were used. The dimensionless parameters Re_b, D^+ and Re_τ fully describe the case of an unladen flow and compare very well to the the experiments of [29] (cf. Tab. 5.1). As expected and discussed in detail below, the particles forming the mobile granular bed started to form distinct patterns during the course of the simulations listed in Tab. 5.3. The re-arrangement of particles, however, may locally lead to a change of the channel height ultimately increasing the Froude number $Fr = U_b/\sqrt{gH}$. Under experimental conditions, this increase can eventually result in surface waves, which is an unwanted feature for precise measurements. Moreover, the numerical setup of a rigid-lid boundary does not allow to display surface waves, either. The requirement of a constant channel height was checked by comparing the parameters investigated with experimental setups of low submergence. Manes *et al.* [99] investigated experimentally the range of Froude numbers for which surface undulations in the present type of setting can be disregarded. They found this to be the case up to $Fr = 0.56$ obtained with a submergence as low as $H/D = 2.5$. With the Froude number being $Fr = 0.35$ for the present conditions, the parameter range used is safely away from any problem of this kind. Thus, the rigid-lid boundary condition applied is well justified even for particle clusters locally.reducing the channel height down to $H/D = 4$. Choosing the fluid to be water with a bulk velocity of $U_b = 0.21\,m/s$ as used in [99] yields a flow depth of $14\,mm$ and the particle diameter of $1.6\,mm$ would correspond to coarse sand. The Froude number would be $Fr = 0.56$. The parameter values of the present study are similar to the range realized in the study [96], which investigated single trajectories of particles with a diameter $D = 1.15\,mm$ in a channel with H/D varying between 7 and 10. Hence, the present setup allows for investigation of fluid–particle interaction in the transitionally rough regime.

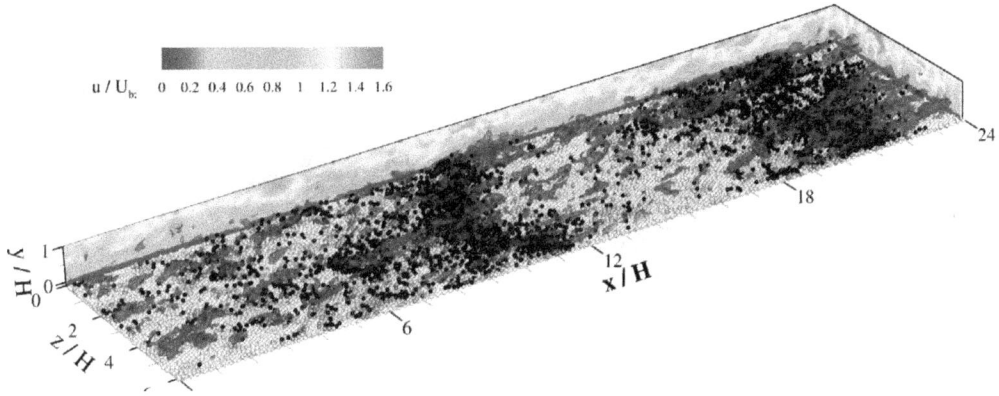

Figure 5.5: *Instantaneous particle distribution of the reference run (*Ref). *Contour plots on the sides of the domain as for Fig. 5.1, 3d-iso-surfaces of fluid fluctuations with* $u'/U_b = -0.3$ *in blue. Particle colors: gray = fixed, white =* $|u_p| < 1.5u_\tau$, *black =* $|u_p| > 1.5u_\tau$.

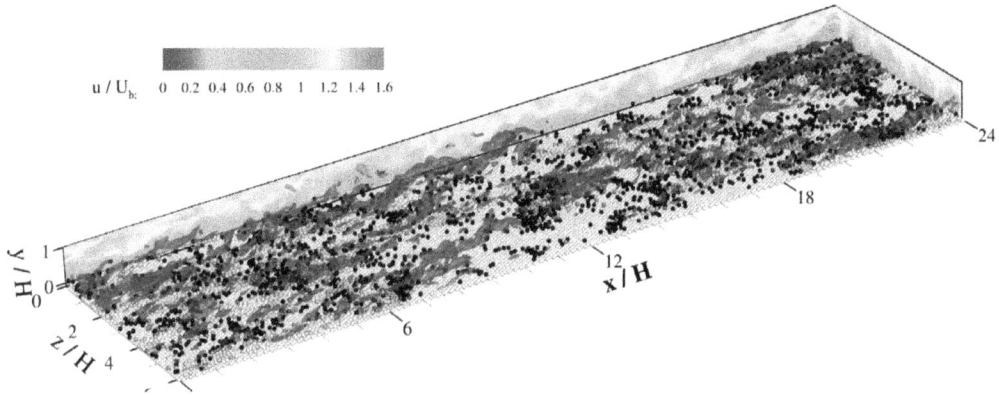

Figure 5.6: *Same as Fig. 5.5, but for case* FewPart.

5.3   Fluid–particle interaction

5.3.1   Pattern formation

The particle patterns which developed in the five simulations are illustrated in Fig. 5.5 – 5.9. The reference run (case *Ref*) features two large-scale structures with their major axis oriented in spanwise direction that move across a layer of inactive particles and have a distance of about $12H$ (Fig. 5.5). These remind of a dune-like roughness feature as discussed, e.g., in [9, 47]. Reducing the mass loading for case *FewPart* yields streamwise oriented clusters of mostly inactive particles of considerable streamwise extent (Fig. 5.6). In between these structures, small-scale clusters of moving particles travel with high velocity along the troughs. This pattern is similar to ridges described in [110, 142] and Fig. 4.2.

Although real dunes are usually observed with large submergences and large numbers of particles, mostly resting over considerable time intervals, the term "dune" is used here as a metaphor to address the patterns observed in the present simulations. A similar precaution

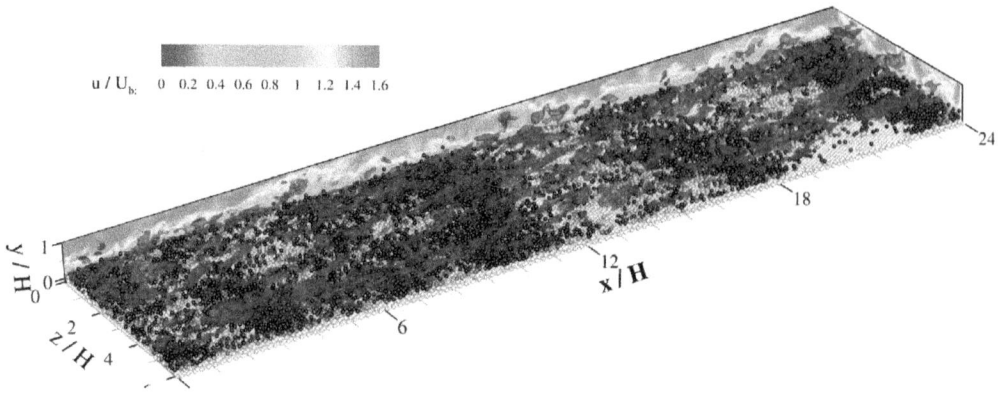

Figure 5.7: *Same as Fig. 5.5, but for case* ManyPart.

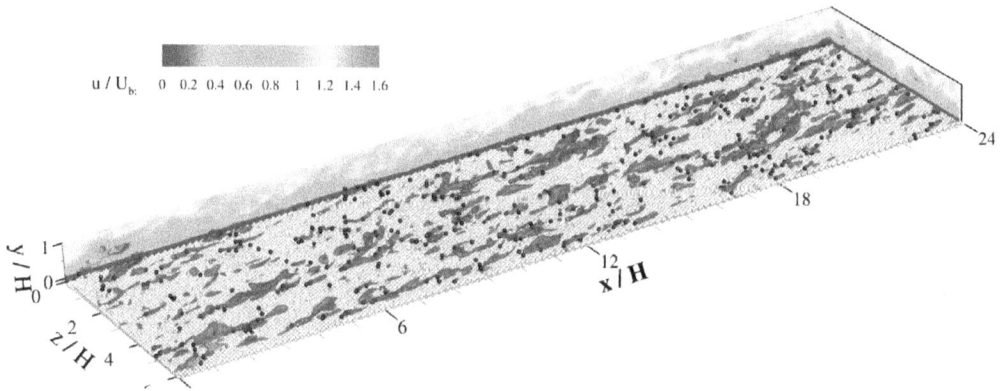

Figure 5.8: *Same as Fig. 5.5, but for case* LowSh.

applies to the designation of ridges.

A higher mass loading enhances the development of the dune-like structures observed in case *Ref* (case *ManyPart* in Fig. 5.7). The structures travel across a closed layer of resting particles with a spacing of roughly $12H$, although the dunes are not as pronounced as in case *Ref*. The coherent fluid structures indicated by iso-surfaces of negative fluid fluctuations increase in size as the turbulent fluctuations are strongly enhanced by the large amount of mobile particles.

Lowering the mobility with respect to case *Ref* results in an almost close bed of inactive particles. Only a small percentage of particles is eroded and transported across the plane bed (Fig. 5.8). A higher mobility of the particles has a somewhat similar impact on the flow as observed in case *ManyPart* (case *HighSh* in Fig. 5.9). Although the mass loading is lower, the percentage of eroded particles is larger than observed in *Ref*, which enhances the development of dune-like structures. The dunes are less pronounced in this case compared to *Ref* and *ManyPart*. The larger amount of mobile particles accelerates the outer flow.

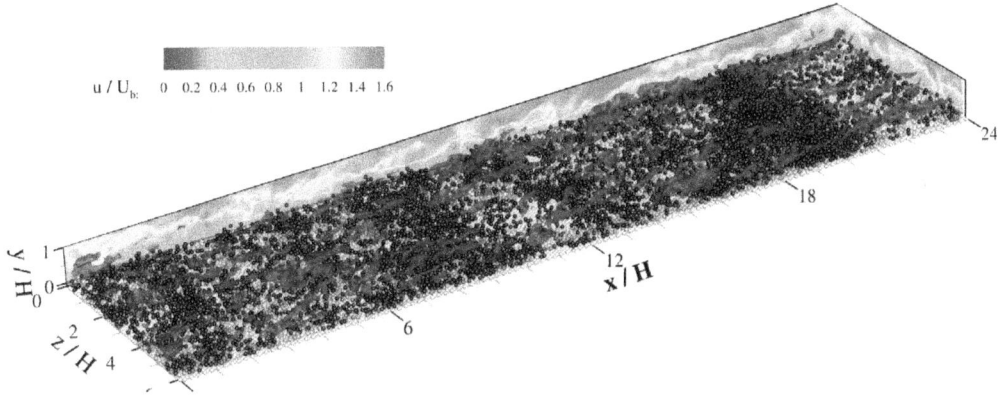

Figure 5.9: *Same as Fig. 5.5, but for case* HighSh.

5.3.2   Modification of fluid properties

The above observations immediately show that the five simulations yield different types of bed-load layers. The resulting particle patterns and fluid behavior is now analyzed in a statistical sense. The hydraulic resistance, for example, is related to the porosity of the disperse phase [112], defined as

$$\phi_{AT}(y) = \langle\gamma\rangle_{x,z,t}(y) \tag{5.1}$$

with the average defined as

$$\langle\gamma\rangle_{x,z,t} = \frac{1}{L_x\,L_z\,t_{aver}} \int\limits_{0}^{L_x}\int\limits_{0}^{L_z}\int\limits_{t_{init}}^{t_{init}+t_{aver}} \gamma(x,y,z,t)\mathrm{d}t\,\mathrm{d}z\,\mathrm{d}x \quad, \tag{5.2}$$

where $L_x L_z$ is the area of a horizontal plane of the domain, t_{init} the simulation time to reach the state of equilibrium as discussed in App. C, and γ an indicator function equal to 1 in the fluid and 0 otherwise. Note that this definition is consistent with the global averaging procedure of the total porosity ϕ_{VT} as described in Sec. 4.4.4. In this chapter, the definition of coordinate-wise averaging is given by the equality in (5.2), while the indices of quantity on the left hand side indicate the operator of the global averaging procedure in streamwise and spanwise direction and time. Thus, the porosity is a measure to quantify the amount of fluid displaced by the disperse phase and is naturally obtained with the present particle-resolving approach. Similar to the porosity, the fluid property θ averaged over a volume occupied by fluid only, can be defined as $\langle\theta\rangle(y) = \langle\theta\,\gamma\rangle_{x,z,t}$. This averaging procedure was applied using a sampling interval of roughly one bulk unit.

The resulting wall-normal profiles of mean quantities are shown in Fig. 5.10 – 5.12. As expected, the mean fluid velocity (Fig. 5.10b) reflects the distribution of porosity over depth (Fig. 5.10a). The deceleration due to the transported particles is compensated by an acceleration of the outer flow. The profiles of porosity obtained for the different cases collapse below $y \approx -0.1H$ since the lowest plane of particles constituting the rough bed is fixed in all cases. Above this layer, a second peak at $y \approx -0.055H = -0.5D$ is observed for all cases. Since this corresponds to particles located in the pockets of the underlying rough bed, these must be resting. As supported by the profile of $\langle u_p\rangle$ shown in Fig. 5.13 below, more than 26% of the mobile particles are moving on top of the fixed bed in the reference

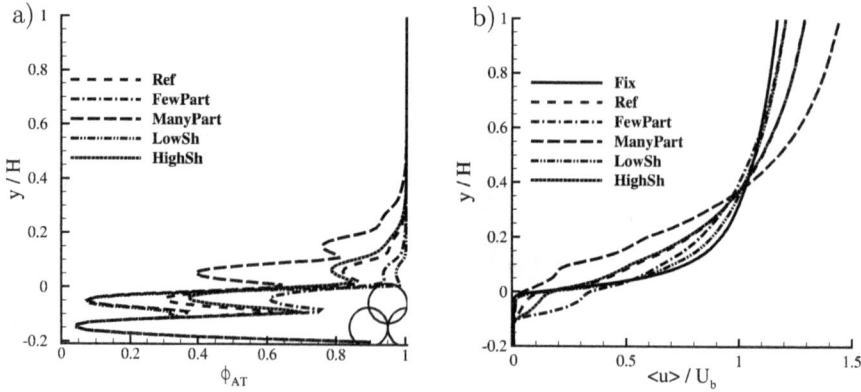

Figure 5.10: *Wall-normal profiles of mean quantities. a) Average particle positions in terms of porosity b) Mean streamwise fluid velocity.*

case, forming a bed-load layer up to $y = 0.3H$ with the dune-like structures discussed before. Reducing the mass loading also reduces the amount of moving particles to 16%, i.e. 8% of a full layer. The inactive ridges have a sizable hiding and shading potential where particles can be entrapped easily [101]. Increasing the mass-loading yields a closed layer (97% of a full layer) of resting particles on top of the fixed bed. Furthermore, a second layer of resting particles forms at $0 < y/H < 0.11H$ with a similar density as the one observed for case *Ref*. The active particles form a bed-load layer on top of these two layers of resting particles.

In case *LowSh*, the resting particles form an almost entirely closed layer, which is very similar to the lower layer reported for case *ManyPart*. The remaining 3% of mobile particles are fully exposed to large fluid stresses and the resulting drag keeps them in motion with a high frequency of collision until a gap in the inactive layer is found. Due to the small amount of moving particles, the mean fluid velocity profiles of *FewPart* and *LowSh* collapse in the outer flow, while the profile of case *Ref* shows large deviations from the unladen case over the entire channel. Increasing the mobility leads to a similar void distribution as reported for case *Ref*, albeit the amount of active particles has increased in this case.

The turbulent fluctuations of the flow are defined as $\theta'(x, y, z, t) = \theta(x, y, z, t) - \langle\theta\rangle(y)$, where θ is any of the velocity components. The resulting Reynolds stresses are displayed in Fig. 5.11. At the free surface, they exhibit the classical pattern related to energy transfer from vertical fluctuations to tangential ones as described, e.g., in [110]. This is not further discussed here, as the focus is on the influence of active and inactive particles concentrated at the bottom. It is noted that the Reynolds stresses are enhanced by the mobile particles in all five cases. Obviously, the Reynolds shear stress, $\langle u'v'\rangle$, as well as the normal stresses induced by the wall-normal component, $\langle v'v'\rangle$, and the spanwise component, $\langle w'w'\rangle$, increase with decreasing porosity in the present cases. For case *Ref*, the streamwise fluctuations have their maximum at $y \approx 0.11H = 1D$, i.e. at a position slightly above the one with entirely immobile particles (case *Fix*) at $y \approx 0.75D$. Furthermore, the maximum of $\langle u'u'\rangle$ is almost doubled from *Fix* to *Ref*. The other Reynolds stresses are increased between the two cases by an even larger factor, with the maximum shifted to slightly higher elevations. This effect is even more dominant for case *ManyPart*, where the peak of $\langle u'u'\rangle$ is increased by another 18 % with respect to case *Ref*. In this case the wall-normal position of the maximum has shifted to $y = 0.22H = 2D$, since below this elevation, particles are mostly resting. In case *FewPart*,

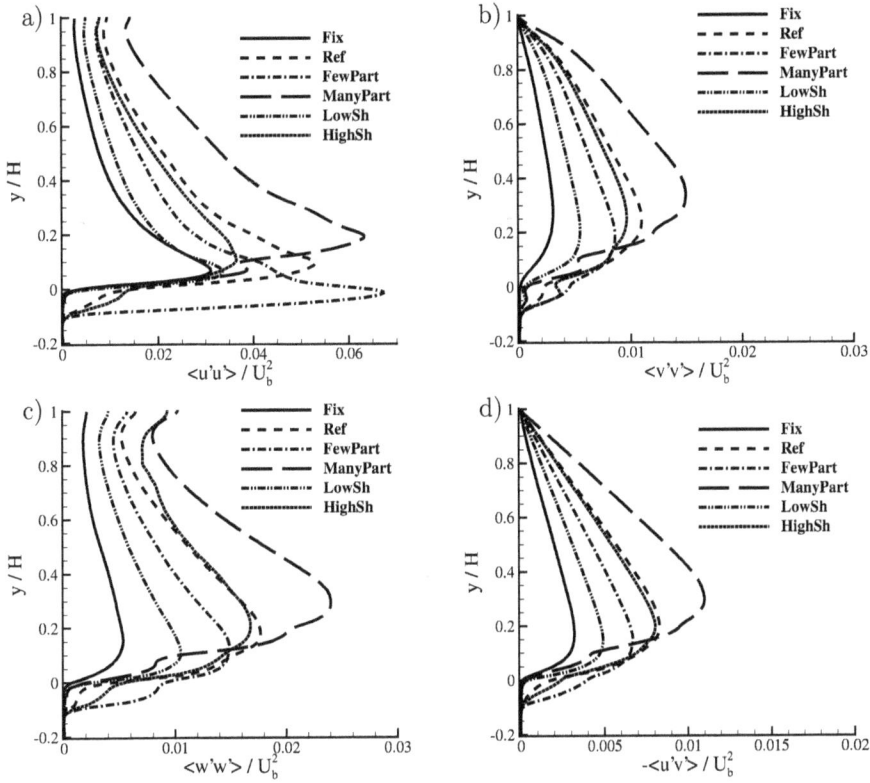

Figure 5.11: *Wall-normal profiles of normalized fluid Reynolds stress. a) Streamwise component* $\langle u'u'\rangle/U_b^2$, *b) wall-normal component* $\langle v'v'\rangle/U_b^2$, *c) spanwise component* $\langle w'w'\rangle/U_b^2$, *d) shear stress* $-\langle u'v'\rangle/U_b^2$. *Note that the scales are identical for b) – d) to allow comparison of the different cases, while the scale for* $\langle u'u'\rangle$ *covers a wider range.*

the streamwise component $\langle u'u'\rangle$ has a peak which occurs at $y = -0.1H$, a lower elevation as for case *Ref*, due to the smaller number of particles in that simulation. Noteworthy is the increased value of the maximum compared to case *Ref*, which is due to the formation of the streamwise ridges. The ridges introduce heterogeneity in spanwise direction and form pronounced troughs. In these troughs the fluid velocity is larger than in other cases, as seen in Fig. 5.10b, and the effective Reynolds number built with the effective submergence is also somewhat larger than in the reference case. Furthermore, relatively fast particles propagate in these troughs as illustrated by Fig. 5.13a and discussed below.

The increase of $\langle u'u'\rangle$ from case *Fix* to case *Ref* results from the spatial heterogeneity of the particle volume fraction as well as from truly turbulent motion. The other Reynolds stresses, in contrast, are smaller for case *FewPart* compared to case *Ref*. Finally, only little alteration of $\langle u'u'\rangle$ with respect to the case with immobile sediment is observed in case *LowSh*. This is due to the fact that only few particles are mobilized by the flow. The relative impact on the other stresses, however, is larger, as they increase by factors of about 1.5 to 2.0 between case *Fix* and *LowSh*. For case *HighSh*, the Reynolds stress, $\langle u'v'\rangle$, $\langle v'v'\rangle$, $\langle w'w'\rangle$ have the same magnitude as the the stresses reported for case *Ref* with their peak value at the same elevation but slightly below this reference. The normal stress $\langle u'u'\rangle$, however, has decreased substantially by more than 34 % with respect to case *Ref*. This effect can be attributed to

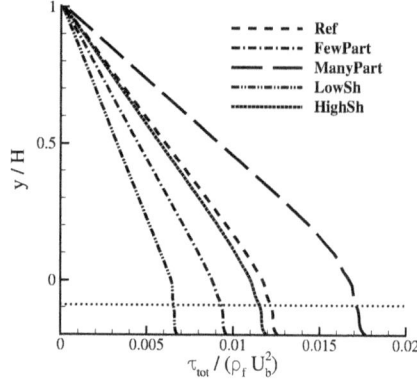

Figure 5.12: *Wall-normal profile of the total shear. The dotted line indicates the limit of the layer of fixed particles constituting the rough wall.*

Case	$\langle f_x \rangle_t H/U_b^2$	$\langle f_x' f_x' \rangle_t H^2/U_b^4$	$\tilde{\tau}_w/(\rho_f U_b^2)$	u_τ/U_b	D^+
Ref	$1.186 \cdot 10^{-2}$	$0.177 \cdot 10^{-2}$	$1.222 \cdot 10^{-2}$	0.101	33.2
FewPart	$0.856 \cdot 10^{-2}$	$0.201 \cdot 10^{-2}$	$0.935 \cdot 10^{-2}$	0.089	29.0
ManyPart	$1.780 \cdot 10^{-2}$	$0.187 \cdot 10^{-2}$	$1.716 \cdot 10^{-2}$	0.120	39.3
LowSh	$0.641 \cdot 10^{-2}$	$0.105 \cdot 10^{-2}$	$0.657 \cdot 10^{-2}$	0.074	24.3
HighSh	$1.123 \cdot 10^{-2}$	$0.175 \cdot 10^{-2}$	$1.157 \cdot 10^{-2}$	0.107	32.3

Table 5.4:   *Time-averaged value of the volume force f_x and its standard variation as well as the resulting bottom friction extracted from Fig 5.12 and the corresponding dimensionless quantities.*

the higher mobility which results in a lower slip velocity $\langle u \rangle - \langle u_p \rangle$ as demonstrated below. The increase of the turbulent fluctuations observed for the present simulations is in line with the numerical results reported by Shao *et al.* [140]. An increase of the Reynolds shear stress goes along with an increase in hydraulic roughness. Defining a wall-shear stress is already delicate for rough walls (see, e.g. [34] or [116]). If, in addition, mobile particles are present in the flow, this becomes even more challenging. Here we apply an approach proposed in [116] by defining a total shear stress $\tau_{tot}(y)$, comprising all exchange of momentum at a given elevation y, via a momentum balance

$$\tau_{tot}(y) = \rho_f \langle f_x \rangle_t \int_y^H \phi_{AT}(y^*) \mathrm{d}y^* \quad . \tag{5.3}$$

The integral on the right hand side represents the momentum introduced by the driving volume force f_x, which is balanced by the total shear. In other words, τ_{tot} is defined by equation (5.3). Choosing a suitable position y_w to define an average position of a "wall" in or below the bed-load layer allows to determine $\tilde{\tau}_w = \tau_{tot}(y_w)$. Here, $y_w = -0.82D = -0.09H$ is chosen, which is the position of the top of the fixed particles (black circles in Fig. 5.2 and horizontal dotted line in Fig. 5.12).

The resulting values are reported in Tab. 5.4. Compared to the bottom shear stress in case *Fix*, $\tau_w = 0.513 \cdot 10^{-2}$, the increase in resistance is largest for case *Ref*. The resulting

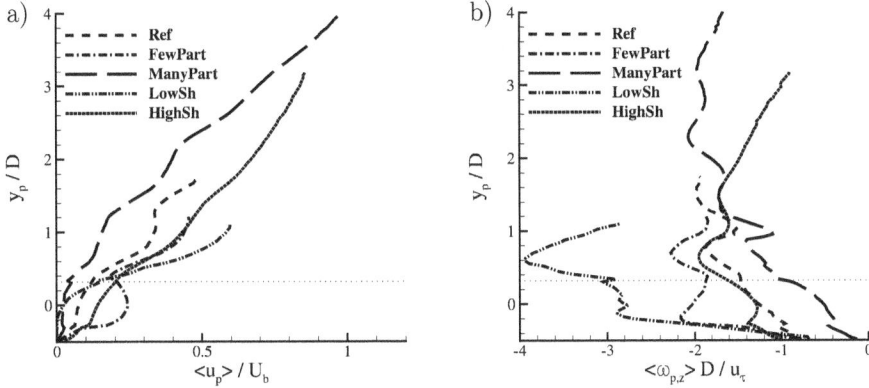

Figure 5.13: *Wall-normal profiles of mean quantities describing the movement of the mobile particles. a) Mean streamwise velocity component, b) mean rotation rate of the particles in spanwise direction. The vertical coordinate is the normalized position of the particle center, the horizontal dotted line is located at $y_p/D = 0.32$.*

increase in dimensionless hydraulic roughness from $D^+ = 21$ to $D^+ = 33$ is by more than 50%. Case *FewPart* contains only half a layer of mobile particles. Nevertheless, the resulting resistance is still increased by 38% over case *Fix*. An increased mass-loading enhances the hydraulic resistance by up to 77 % (case *ManyPart*). As observed above for other quantities, case *LowSh* exhibits only little difference with respect to case *Fix*. Here, this results in an increase of D^+ by only 16%, while the modified particle Reynolds number reported for case *HighSh* is similar to case *Ref*. Comparing the the cases with a substantial amount of active particles (cases *Ref*, *ManyPart*, and *HighSh*) with the unladen flow or the dilute regime of case *LowSh* illustrates that mobile sediment can indeed yield increased resistance also observed by Shao *et al.* [140] as mentioned above.

5.3.3   Modification of particle properties

Particle statistics were gathered every fluid time step (i.e. $2 \cdot 10^{-3} H/U_b$). In accordance with the definition given in (5.2), the operator $\langle \theta_p \rangle(y) = \langle \theta_p \rangle_{x,z,t}$ is used, here. Since the relative density is close to or even below the threshold of motion predicted by the Shields criterion, particles tend to stay in the vicinity of the bed. To ensure a sufficient quality of the average, results are reported only for converged bins, defined by the condition that increasing the averaging time by $24H/U_b$ changes the value of the bin by less than 1%. Nevertheless, the data shown represent more than 98.5% of the samples.

The wall-normal profile of the particle velocity is illustrated in Fig. 5.13a. For case *Ref*, the particle velocity increases from the bottom up to $y/D = 1$. This interval represents the bed-load layer dominated by permanent frictional contact of the active particles with the resting and fixed ones. Large clusters are eroded through the dune-like structures. The particle interaction leads to a local minimum of the particle velocity at $y/D = 1.32$. Above this height, the regime becomes dilute and the impact of particle interaction decreases. In case *FewPart*, the wall-normal profile of the mean translational particle velocity shows high positive values for $y < 0$. This is caused by small-scale particle clusters that travel in the troughs with a rather high translational velocity, observed in animations of the particle motion of which Fig. 5.6 presents a still. The kinetic energy transported by these clusters can be transferred to other particles by collisions leading to the temporary erosion of the ridges at different locations. Above this region, the enhanced hiding and shading mechanism and the frictional contact with the inactive ridges decelerate the particles up to $y_p = 0.35D$. As soon as particles reach the outer flow, particle interaction decreases and the translational velocity approaches the fluid velocity. Smaller values for particle velocity are reported for case *ManyPart* as very high volume fractions are observed in the near-wall region. The particle-particle interaction decreases their velocity. Since particles reach bigger heights in the channel, the thickness of the bed-load layer increases as well.

To be eroded, the heavy particles of case *LowSh* embedded in the plane bed have to overcome a difference in height of $0.32D$, generated by the saddle between one pocket and a neighboring one. A particle resting in such a pocket has the wall-normal coordinate $y_p = 0.32D$ as illustrated by the upper right particle in Fig. 5.2b. Therefore, starting from the bottom, the particle velocity increases continuously and has a small kink at $y_p = 0.32D$. Above this height, the statistical data predominantly represent the particles with higher kinetic energy already eroded. This kink is not present for *Ref*, as larger clusters are eroded through the dune-like structures, while in *LowSh* the erosive events mobilize single particles only. The increase of the particle velocity in *Ref* is smaller compared to case *FewPart* and *LowSh*. The particles exhibit strong frictional forces by collision as they constitute dunes with strong particle-particle interaction through collisions as well as hiding and shading. This effect is not present for case *HighSh* since the gross part of the particles are active and transfer of kinetic energy between particles by collisions happens very frequently decreasing the probability of large scale clusters of resting particles.

The profiles of the angular velocity in Fig. 5.13b show that the particles of *Ref* and *ManyPart* have a lower rotation rate than those of the other cases. Since the particles mostly travel in dune-like structures, the particle-particle collisions also happen with a lower relative tangential velocity. Fast moving particles of the other five simulations mostly collide either with the fixed bed (*FewPart*), the inactive ridges (*FewPart*), or the inactive plane bed (*LowSh*). This enhances the rotational motion. Again, a kink at $y_p = 0.32D$ can be seen for *LowSh*. The erosive events represented by bins $-0.2D < y_p < 0.32D$ happen via a rolling motion with a constant rotation rate. Above this height, the statistical data is dominated by moving particles colliding very frequently. These oblique collisions with inactive particles enhance the angular velocity even further.

The averaged particle behavior can be linked to the fluid by defining a particle Reynolds number as

$$Re_p = \frac{|\langle u \rangle_{x,z,t} - \langle u_p \rangle_{x,z,t}|D}{\nu_f} \quad , \tag{5.4}$$

where $|\langle u \rangle_{x,z,t} - \langle u_p \rangle_{x,z,t}|$ is the slip velocity of a particle in a given height with respect to the time-averaged fluid velocity (data shown in Figs. 5.10b and 5.13a). The wall-normal profiles of the five different cases are assembled in Fig. 5.14. It can be noted that in all simulations Re_p is lower than a value of 130. According to [39], this corresponds to a *Steady Wake Regime* for particles exposed to a uniform flow. Hence, the wake created by the particles has a dimension of up to $1D$ in length according the this reference. The same applies for spheres exposed to a wall-bounded flow. As a matter of fact, it was observed by [163] that the closer the distance of a particle to the wall the less the difference of the resulting drag and lift to the uniform flow-scenario addressed above. This explains the high probability of an active particle to be entrapped in the wake of a resting particle, a mechanism contributing to the

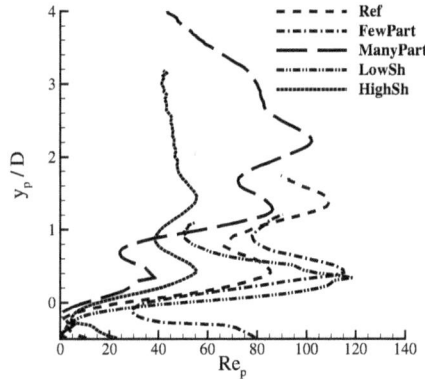

Figure 5.14: *Wall-normal profiles of particle Reynolds number based on the slip velocity $|\langle u \rangle - \langle u_p \rangle|$ of the mobile particles.*

development of ridges (case *FewPart*). The relatively small extent of the wake regions also shows that particle-particle interaction of active particles is mostly triggered by collisional processes rather than by long-range interactions due to the pressure field.

The local maxima of the wall-normal profiles of Re_p illustrate the particle interaction imposed by the strong geometrical ordering of the fixed bed on the bottom of the channel. For case *Ref*, maxima are reported for $y_p = 0.4D$ and $y_p = 1.4D$. The corresponding values are $Re_p(y_p = 0.4D) = 85$ and $Re_p(y_p = 1.4D) = 109$, respectively. A peak in Re_p corresponds to wall-normal positions of a layer of the hexagonal packing. At these heights, the interaction between fast active and resting particles is strongest slowing down active particles by frictional forces. The value at $y_p = 1.4D$ is higher, because at this height, the exposure of the particles to the turbulent flow has increased. In addition to the maxima discussed for case *Ref*, a maximum at $y_p = -0.5D$ is reported for case *FewPart*. Since at this height, no closed bed of resting particles is reported, regions of high-speed fluid develop in the troughs between the ridges. The slip velocity is even amplified by more than 35 % with respect to case *Ref*. For case *ManyPart*, two closed layers of resting particles are reported with a mean wall-normal position of $y_p = -0.5D$ and $y_p = 0.32D$. Hence, Re_p is even below the nominal value of a steady wake regime. The values of the local maxima in the outer flow are very similar to the values reported for case *Ref*, although shifted by $1D$ in positive wall-normal direction.

For case *LowSh* a closed bed of resting particles was reported in Sec. 5.3.2. The data reported for bins $-0.5D < y_p < 0.32D$, hence represent particles that are eroded and lifted out of the closed bed. This mechanism is addressed in detail in Sec. 6.2 below. At this interval an increase of Re_p is observed with its peak value of 120 at $y_p \approx 0.32D$. Above this height, the particle Reynolds number decreases sharply to a value of 50 due to the missing interaction between resting and active particles that mostly travel in a saltating motion. Among the cases investigated, case *HighSh* shows the lowest values of Re_p. This was expected, since the high mobility leads to low inertia and particles can follow the surrounding fluid more easily. Above $y_p = 1.8D$, where particle-particle interaction becomes negligible, no distinct local maxima can be reported, but a linear decrease of Re_p with increasing height is observed.

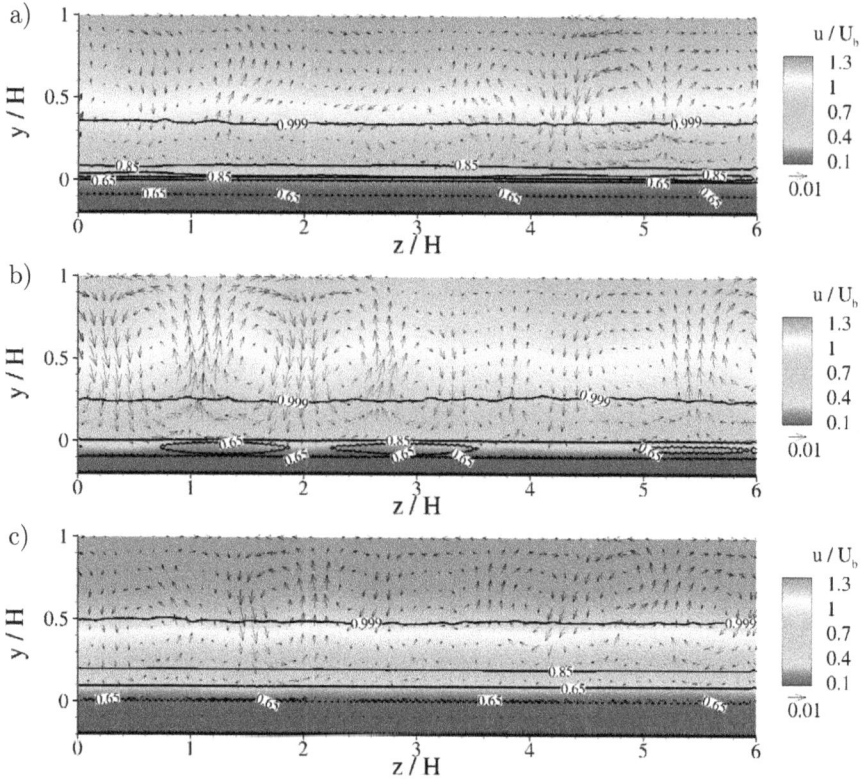

Figure 5.15: *Two-dimensional map of the streamwise fluid velocity component illustrated by the contour plot of the mean streamwise velocity. The wall normal and spanwise fluid velocity are illustrated by $(\langle v \rangle_{x,t} - \langle v \rangle_{x,z,t}, \langle w \rangle_{x,t} - \langle w \rangle_{x,z,t})^T$, and the porosity is illustrated by iso-lines. a) Case Ref, b) case FewPart, c) case ManyPart. Axes are not to scale.*

5.3.4   Length scales of particle clusters and coherent fluid structures

For the five cases with mobile particles, the average structure of the layer of resting particles and the bed-load layer is elucidated in more detail by the two-dimensional maps showing iso-lines of the porosity $\phi_{VT}(y, z) = \langle \gamma \rangle_{x,t}$ in Figs. 5.15 and 5.16. Moreover, contours of the streamwise velocity component as well as vector plots of the wall-normal and spanwise component are shown with similar definition as given in Sec. 4.5.5. Here, the components are defined as

$$v_{sec} = \langle v \rangle_{x,t} - \langle v \rangle_{x,z,t} \tag{5.5a}$$

$$w_{sec} = \langle w \rangle_{x,t} - \langle w \rangle_{x,z,t} \tag{5.5b}$$

to illustrate the magnitude of the secondary currents. The iso-lines of $\phi_{VT}(y, z)$ were chosen to display the particle pattern. In addition, another iso-line of 0.999 was drawn, since a value larger than this threshold characterizes a dilute regime with negligible feedback on the flow according to [12].

The dune-like structures of case *Ref* form a region of high particle concentration in the interval $0 < y < 3.2D$ with a rather homogeneous distribution in spanwise direction (Fig. 5.15a).

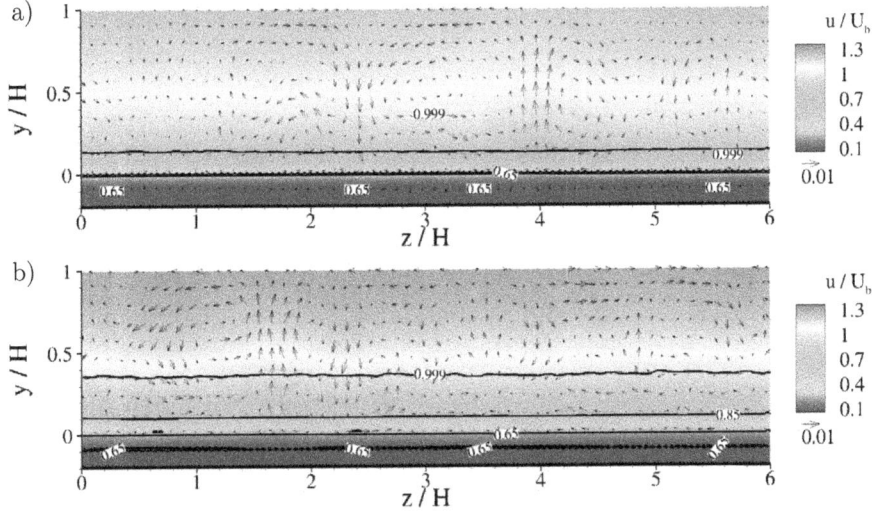

Figure 5.16: *Same as Fig. 5.15 but for a) case* LowSh *and b) case* HighSh. *Axes are not to scale.*

Nevertheless, cells of secondary currents develop with different intensities. This feature is addressed in more detail in Sec. 6.3. In contrast to case *Ref*, distinct spatial heterogeneity is introduced in case *FewPart* by the ridges indicated by the regions of low $\phi_{AT}(y, z)$ at $y/H \approx 1.5, 3.5, 5.5$ (Fig. 5.15b), which evolve very slowly in time. This is the reason, why the spacing of the rigdes of $2H$ witnessed in experiments [110, 142] has not yet fully developed to an equidistant pattern for the averaging time simulated. Moreover, the cited experiments investigated fully inactive particle structures with very stable conditions, which is not the case in the present simulation. Furthermore groups of particles are transported in the troughs. Both effects blur the length scales observed. As expected, the spanwise heterogeneity leads to strong intensities of the secondary currents.

Since the amount of displaced fluid is largest for case *ManyPart* (Fig. 5.15c), the level of $\phi_{VT} = 0.999$ is located at higher elevations than in the other cases, while this level is lowest for case *LowSh*. For this case, only few particles are active, which implies homogeneous conditions for the outer flow. Hence, the intensity of the secondary currents is very low for this case (Fig. 5.16a). Again, strong similarity between case *Ref* and case *HighSh* can be observed (Fig. 5.16b).

In order to quantitatively determine the typical length scales of the particle clusters, a two-dimensional Cartesian Distribution Function (CDF) was defined along the lines presented in [27, 71, 137]

$$G(\xi_x, \xi_z) = \frac{L_x L_z}{\Delta \xi_x \Delta \xi_z} \frac{N_{pairs}(\xi_x, \xi_z)}{N_{p,tot}} \quad . \tag{5.6}$$

The value N_{pairs} is the number of particles at the distance ξ_x in x- and ξ_z in z-direction within an interval $\Delta \xi_x$ and $\Delta \xi_z$, respectively, averaged in space and time. Here, the discretization interval of the CDF was set to $\Delta \xi_x = \Delta \xi_z = 1D$. The total number of possible particle pairs is $N_{p,tot} = N_{p,mob}(N_{p,mob} - 1)$. Roughly speaking, $G(\xi_x, \xi_z)$ represents the probability of a particle at a point x_p, z_p to find a pair particle at distance ξ_x, ξ_z from its position. A uniform value of 1 corresponds to a random particle distribution [137]. Due to the statistical homogeneity of the configuration in x and z, G does not depend on x_p and z_p.

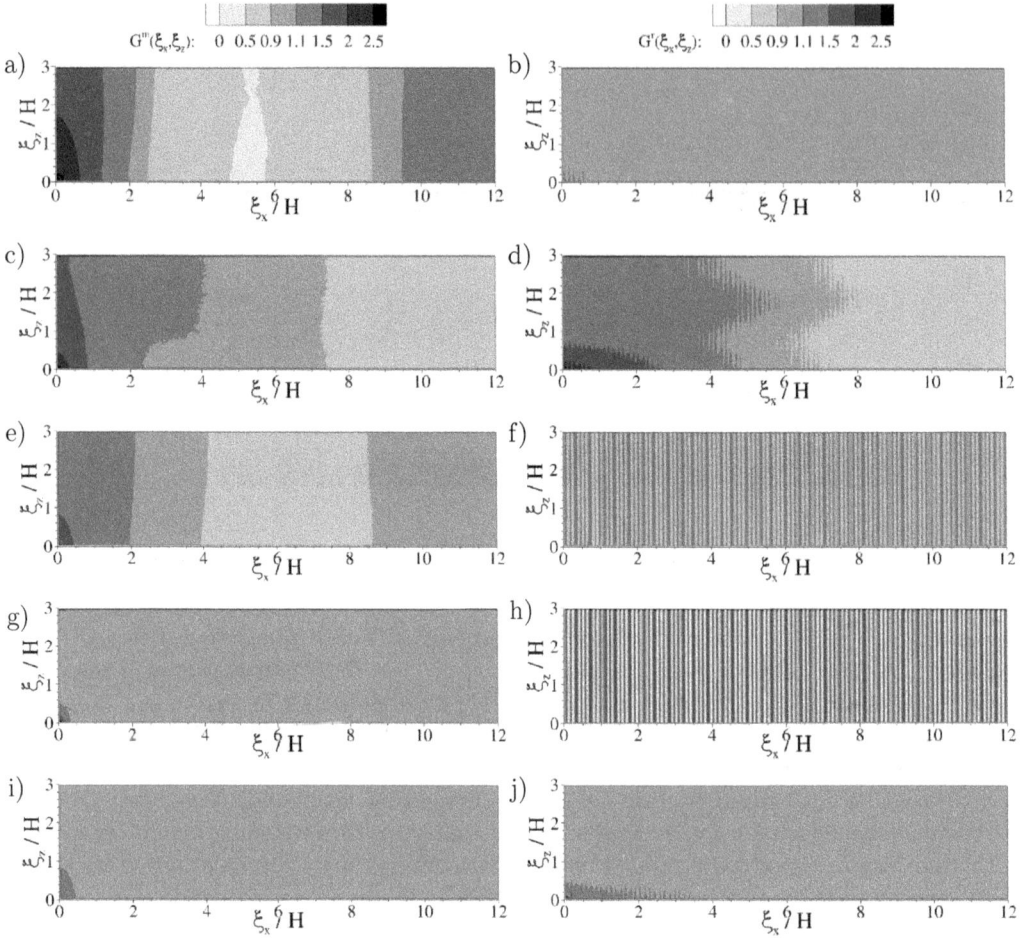

Figure 5.17: *CDF for moving (G^m) and resting (G^r) particles. a) G^m of case Ref, b) G^r of case Ref, c) G^m of case FewPart, d) G^r of case FewPart, e) G^m of case ManyPart,f) G^r of case ManyPart, g) G^m of case LowSh, h) G^r of case LowSh, i) G^m of case HighSh, j) G^r of case HighSh, Due to the symmetry of G in ξ_x and ξ_z, the data is drawn for positive coordinates only.*

Since it was shown in Figs. 5.13a, 5.15, and 5.16 that particles stay in the near-wall region, distances in the wall-normal direction ξ_y were not accounted for and integration over y was performed, so that G is a purely two-dimensional quantity.

To allow physical interpretation, the CDF was, in a next step, conditioned by the particle velocity, with G^m only accounting for pairs of two moving particles, $|\mathbf{u}_p| > u_\tau$, and G^r only pairs of two resting particles $|\mathbf{u}_p| < u_\tau$ and $N_{p,tot}$ computed with only these particles. This measure allows to distinguish between dune-like and small-scale clusters on the one hand, and the closed plane bed and the ridges on the other hand. This analysis was performed for the last 120 bulk units of each simulation with a sampling interval of $0.2H/U_b$. The two-dimensional distributions of both G^m and G^r are reported for all cases in Fig. 5.17. Note that in contrast to [137], a symmetric type of pair correlation function was used, i.e. for every particle all the possible pairs are accounted for. The procedure is similar to the analysis presented in [136], where the small-scale clustering of buoyant particles was investigated in a vertical channel. The CDF of the moving particles, G^m, now presents statistical evidence of what was noted before in a qualitative way by visually inspecting the particle motion in Fig. 5.5. To further compare the magnitude of G for the five cases in a quantitative manner, one-dimensional profiles of G^m and G^r, i.e. cuts along ξ_x and ξ_z, are presented in Fig. 5.18. First of all, dune-like structures constituted of active particles are reported for the cases *Ref* and *ManyPart* in Fig. 5.17a and e. The characteristic streamwise distance of $12H$ of the dune-like structures is reported for these cases and confirmed by the corresponding curve in Fig. 5.18a with the structures being more pronounced for the former. For case *FewPart*, the signal of G^r shows a zero crossing at $6H$ in Fig. 5.18b. As $G(\xi_x, \xi_z)$ is symmetric with respect to the origin by definition, this shows that in the present case, the averaged length of the ridges is $12H$. The two-dimensional plot of CDF in 5.17b shows the spanwise spacing of the ridges of $2H$ known for more stable ridges as discussed above. The dilute particle distribution of *LowSh* is close to the random regime throughout the entire domain. The maximum in the first bin at $\xi_x = 1D$ of the curve in Fig. 5.18a indicates that particles mostly travel individually over the inactive bed, possibly teaming up with a second particle. Similar observations of particle pairing were reported for a vertical channel flow laden with sedimenting [149] and buoyant particles [136]. Both studies attribute the particle behavior to the so-called drafting-effect, i.e. local pressure minima on the sides of the particle determine the preferential local alignment of the particle pair.

Despite the strong similarity between case *Ref* and *HighSh* illustrated and discussed for Figs. 5.10–5.15, a different picture becomes evident when looking at the particle distribution of case *HighSh*. Similar to case *LowSh* the particle distribution is close to the random regime, but in this case the amount of active particles has increased substantially. In contrast to the distinct particle clusters reported for case *Ref*, no ordering can be described in a statistical sense, if particles of high mobility travel in a large number across the fixed bed. Since both simulations show similar hydraulic resistance (Tab. 5.4), particle clusters must occur for case *HighSh* as it is also evident in Fig. 5.9. These clusters, however, are not stable in time as they average out to a random distribution during the course of the simulation.

As the ordering of the resting particles is substantially influenced by the hexagonal packing of the fixed bed, this ordering is reproduced in a staggered manner and visible through the quasi-periodic oscillations of G^r in Fig. 5.18b and d. The number of resting particles is largest for *LowSh*, so that the amplitude of the oscillations is strongest in this case. The intensity of the oscillations decreases for *Ref*, as up to 26% of the particles are eroded. Similar observations hold for case *ManyPart*, and *LowSh* (Fig. 5.17f and h). The oscillations occur in addition to the signal of the ridges evident for case *FewPart* and *HighSh* (Fig. 5.17d and i). Moreover, the values $G^m > 1$ of the spanwise profiles shown in Fig. 5.18c and d illustrate for case *FewPart* that despite the existence of the ridges, particle clustering for both, moving and resting particles is observed in the spanwise direction.

It is now interesting to investigate, if and how the particle patterns are reflected by the two-point correlation of the fluid velocity in streamwise direction

$$R_{uu}(r_x) = \frac{\langle u'(x)\, u'(x+r_x)\rangle_{x,z,t}}{\sqrt{\langle u'(x)u'(x)\rangle_{x,z,t}\langle u'(x+r_x)u'(x+r_x)\rangle_{x,z,t}}} \tag{5.7}$$

shown in Fig. 5.19a and b. The analogous quantity in spanwise direction is displayed in Fig. 5.19c and d. The two-point correlation was calculated for the same data set as used for the CDF. In a addition, comparison to the unladen case is provided by case *Fix*. A spatial grid coarsened by a factor of five with respect to the grid of the simulation was used for this

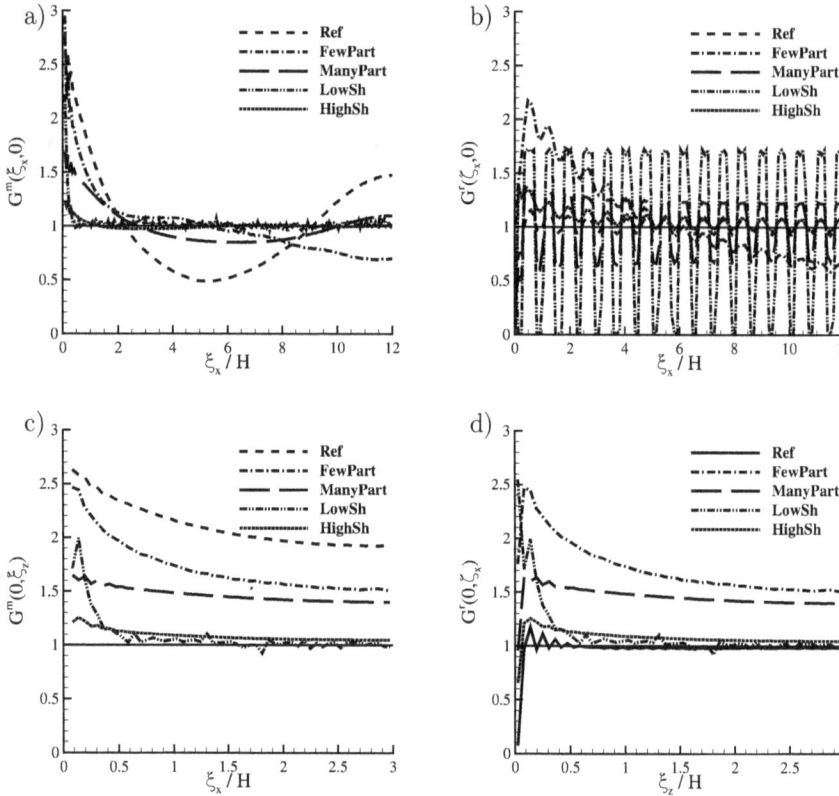

Figure 5.18: *CDF of the disperse phase for moving and resting particles. a) CDF in streamwise direction for moving particles, b) CDF of resting particles in streamwise direction, c) CDF in spanwise direction for moving particles, d) CDF of resting particles spanwise direction.*

purpose. Two wall normal-elevations were selected. The first elevation is close to the layer of resting particles, which is $y = 1.5D = 0.17H$ for case *ManyPart* and $y = 0.5D = 0.056H$ for the five remaining cases. The second elevation is located at $y = 4.1D = 0.46H$, which is the area with reasonably dilute conditions for all cases. The analysis was performed for both elevations.

Starting with the results for case *Fix* it can be recognized that these correspond to the usual behavior of turbulent flow over a rough bed [160, 147]. Streamwise streaks of width $2 \cdot 0.3H = 5.4D$ can be recognized from the pronounced minimum of $R_{uu}(r_z)$ close to the bed in Fig. 5.19c. Fig. 5.19d shows that at higher elevation the typical spanwise size of coherent structures is more than doubled at a distance of $4.1D$ away from the the bed. The streamwise extent of the u-fluctuations is about $6H$ and only slightly larger at higher elevation.

In case *LowSh*, only few particles are mobilized and travel over the bed. The modifications of the fluid structures is relatively small and in the direction of somewhat increased disorder without fundamental changes: the streamwise streaks in $R_{uu}(r_z)$ are of the same width but less pronounced (Fig. 5.19c and d). The streamwise correlation in Fig. 5.19a is smaller by about 50%. The correlation of case *Ref* reflects the substantial motion of the particles. The streamwise correlation $R_{uu}(r_x)$ in Fig. 5.19a has the same shape as the distribution func-

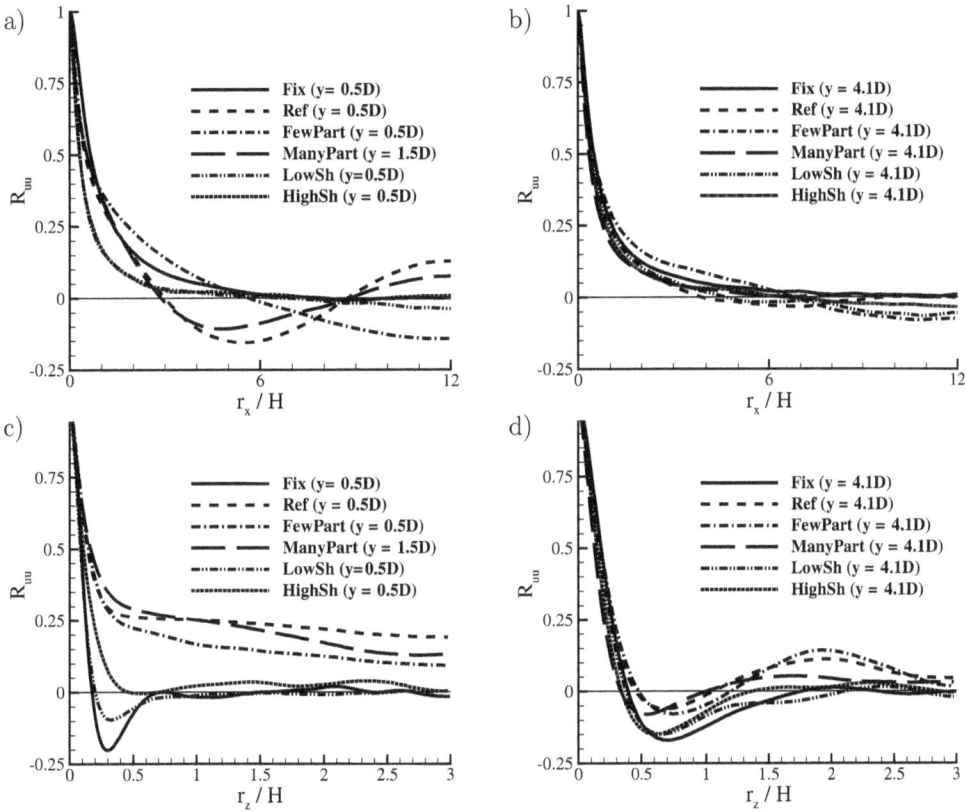

Figure 5.19: *Two-point correlation function of the streamwise fluid velocity component. a) Correlation in streamwise direction in the near-wall region b) correlation in streamwise direction in the dilute region, c) correlation in spanwise direction in the near-wall region, d) correlation in spanwise direction in the dilute region.*

tion of the particles in Fig. 5.18a resulting from the presence of spanwise oriented particle accumulations with a distance of about $12H$. As a result, the fluid velocity is positively correlated over $12H$ and negatively correlated at half this distance. Further away from the bed, at $y = 4.1D$, only very slightly negative correlation remains around $r_x/H = 6$ beyond the peak around the origin. The spanwise correlation $R_{uu}(r_z)$, remains in the interval 0.25 to 0.35 with only slight decay, as a consequence of the spanwise orientation of the particle structures demonstrated by Fig. 5.18c. This feature reflects the secondary currents illustrated in Fig. 5.15a, which occur away from the wall, at $y = 4.1D = 0.46H$, with a spanwise spacing of about $1.4H$ on average. At this distance from the wall no particles are encountered any more (see Fig. 5.10a).

For case *FewPart*, which features streamwise ridges, the statistics of resting particles in Fig. 5.18c and d shows that on average the streamwise particle structures have an extension of about $2 \cdot 6H = 12H$. This is closely matched by the streamwise correlation $R_{uu}(r_x)$ in Fig. 5.19a. For $r_x/H = 0.1...0.5$, these values are the largest among all cases considered. The pronounced minimum at $r_x/H = 12$ shows that the ridges do not extend over the entire domain. In spanwise direction, R_{uu} decreases from 0.25 at $r_z/H = 0.2$ to 0.09 at $r_z/H = 3$. Very pronounced regular ridges would feature a periodicity in spanwise direction of about $2H$

[157], which is not observed for the present case neither in the correlation of the streamwise velocity, nor for the particle statistics shown in Fig. 5.18b and c. This is due to the fact that the ridges have a relatively short length in streamwise direction and that these occur not so regularly all over the domain. The snapshot in Fig. 5.6, for example shows a larger area without resting particles (white) on top of the rough wall (yellow) around $x/H = 12$ and $z/H = 5$. Furthermore, the snapshot features mobile particles (black) traveling in irregular clusters which again disturb any regularity of the fluid streaks. The spanwise correlation for case *FewPart* away from the wall, i.e. above the mobile bed in Fig. 5.19d remarkably shows very much the same behavior as for case *Ref*.

Increasing the amount of mobile particles reveals large-scale fluid structures in the near wall region that have a similar characteristic as the statistics reported for case *Ref* (case *Many-Part* in Fig. 5.19a and c). Their intensity, however, reflected by the minimal and maximal values of R_{uu} is not as strong compared to the values observed for case *Ref*. The cells of secondary currents have a smaller spanwise extent of $z/H \approx 1.7$ as indicated by the peak value in Fig. 5.19d. For case *HighSh*, no large scale vortex organization is visible in R_{uu}. As already mentioned for the particle correlation, this is due to the rather random distribution, which results in an uncorrelated flow field in streamwise as well as spanwise direction for both elevations investigated.

5.4   Conclusions

Highly-resolved simulations of particle-laden flows across an idealized sediment bed constituted of a layer of fixed spheres were performed in the transitionally rough regime using an Immersed Boundary Method and a sophisticated collision model. Variation of the physical parameters, such as mass loading and mobility, lead to five substantially different particle patterns. The presence of mobile particles alters the flow field substantially enhancing turbulence and interacting with the coherent fluid structures. Five cases were reported, dominated by individual mobile particles, spanwise particle clusters and streamwise ridges. The integral impact of the particles on the turbulent channel flow was addressed by means of a momentum balance ultimately quantifying the hydraulic resistance of the mobile bed on the flow. The resulting interaction of the two phases was analyzed employing suitable statistical tools to distinguish between the length scales of eroded and inactive clusters. The length scales for the dominant particle patterns were derived from this analysis and found to match with the length scales of the coherent fluid structures. Even for small scale clusters, a significant effect of the disperse phase on the flow was obtained. The results agree qualitatively with observations from experiments obtained at higher Reynolds number and allow a detailed analysis of the flow. Hence, this approach has proven to be suitable to perform simulations on this scale and in the future may serve as a tool to improve parameters of the classical models used in engineering practice. The simulations build a valuable database for the analysis of distinct flow patterns, of which Chap. 6 addresses two: the flow conditions at erosion and incipient motion events and the flow conditions around the large-scale clusters that travel with a considerable velocity in streamwise direction.

6 Investigation of flow events by conditional averaging

6.1 Introduction

The statistical tools presented so far have proven to give valuable information to describe the physics of uniform, steady open-channel flows as well as the mutual interaction between the two phases. This is especially true for conditions, where a separation between morphological and turbulent time scales takes place. If, on the other hand, one is interested in particular events, this description is not sufficient anymore, because these flow events are usually characterized to be unsteady and highly three-dimensional. This is especially true when dealing with such complex systems like the case of bed-load transport, where singular events can either lead to a disturbance of the flow or act as stabilizing mechanisms. It is well known, that coherent structures dominate wall-bounded flows. In the case of a turbulent flow over a rough wall, coherent structures account for the gross of the energy available for sediment erosion [6, 7]. An analysis employing double-averaging operators across the periodic streamwise and spanwise directions, however, would average out the distinct characteristics of these flow structures. In order to capture local heterogeneity of flow events, conditional averaging is a powerful tool to separate the mechanisms of interest from the general time- and space-averaged behavior of the mean flow [20]. The large domain employed for the simulations presented in Chap. 5 allows for a high realism and, hence, constitutes an ideal data base to perform such conditional averaging procedures.

Two examples of these mechanisms are discussed in the present chapter. First, Sec. 6.2 presents a conditional analysis of the characteristic flow events that lead to particle erosion. This was done using the data from case *LowSh*. Most of the results presented in this section was a joint work elaborated in the Master thesis of Ramandeep Jain [74], which constitutes the basis for a manuscript presented at ETMM10 in 2014 [152]. The criterion for conditioning was incipient particle motion, which gives rise to an ensemble average of independent samples. This study was motivated by the fact, that despite the long history of research on incipient motion [138], the formulae to predict erosion events and incipient motion of inertial particles are of low predictive power yielding errors that well exceed 100% [18]. This may be explained by the empirical nature of these formulae among which the Shields parameter is the most famous [141]. Typically, an erosion event is also associated with a certain duration of large stresses [33, 48]. Another issue, which has reviewed much less attention so far, is the role of mobile particles as a trigger for erosion events.

The second issue, which is analyzed in Sec. 6.3, addresses the characteristic flow field around particle clusters that propagate with a considerable velocity in streamwise direction such as the ones reported by [47]. In particular, the section focuses on the dune-like clusters observed

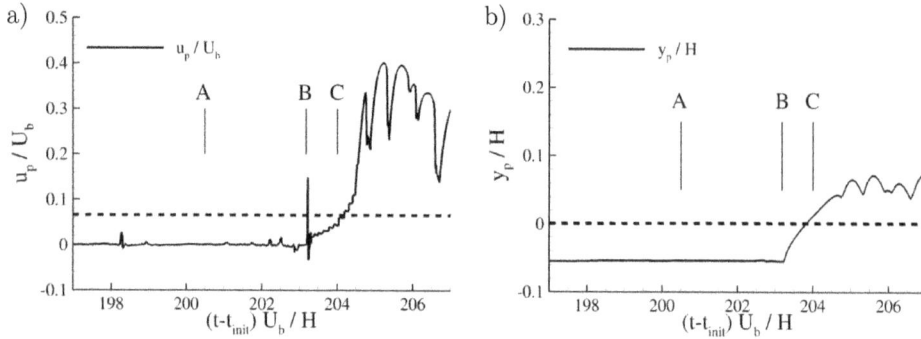

Figure 6.1: *Example of an erosion event. Three representative instants in time are marked with capital letters and correspond to the situations shown in Fig. 6.2 below. Note that instant B corresponds to the nominal start of the erosion $t = t_e$. The dashed horizontal line indicates the criterion to detect incipient motion. a) instantaneous particle velocity, b) wall-normal particle coordinate.*

for case *Ref.* For this configuration, a statistical analysis as proposed in Chap. 4 would completely blur the observed particle patterns. This is remedied by means of a moving frame analysis. Here, the coordinate system was conditioned by a transformation such that the observations are as stationary as possible [64]. Since the time-scales for particle clusters to evolve in time is known to be quite large [9], the simulation time interval was significantly extended to allow investigation of the long-term behaviour of the particle clusters.

6.2   Incipient motion

6.2.1   Identification of erosion events

The potential of a flow to erode a sediment bed is classically assessed by the Shields number $Sh = u_\tau^2/(\rho' g D)$ in the sense that erosion is expected for Sh being larger than some critical value Sh_{crit} depending on the Reynolds number. Here, g is the gravitational acceleration and $\rho' = (\rho_p - \rho_f)/\rho_f$ the relative submerged particle density with ρ_p the particle density and ρ_f the fluid density. To gather proper statistics of single-particle erosion events, a flow laden with particles heavy enough to settle onto the bottom but light enough to be eroded every now and then is needed. This condition is met for case *LowSh* reported in Chap. 5 by choosing $\rho' = 0.182$ resulting in a value 25% lower than the critical threshold of incipient motion according to Shields (1936), which is $Sh_{crit} = 0.034$ for the present conditions.

This simulation was run for a total averaging period of duration $T_{aver} = 297H/U_b$ to detect erosion events with erosion and deposition rates of the disperse phase being in equilibrium. It was reported in Sec. 5.3 that about 3% of the mobile particles are traveling with a high velocity across a layer of temporally resting particles (Fig. 5.8), which is the desired situation for the present study.

A convenient way to detect erosion events for the present scenario is to record the wall-normal position of a particle center, y_p, together with the particle velocity in streamwise direction, u_p. This is exemplified in Fig. 6.1. Particles embedded within the inactive layer at the bottom of the channel have a wall-normal coordinate of $y_p = -0.5D \approx -0.055H$ and a velocity of $u_p \approx 0$ (situation shown in Fig. 6.2a). An erosion event leads to a substantial

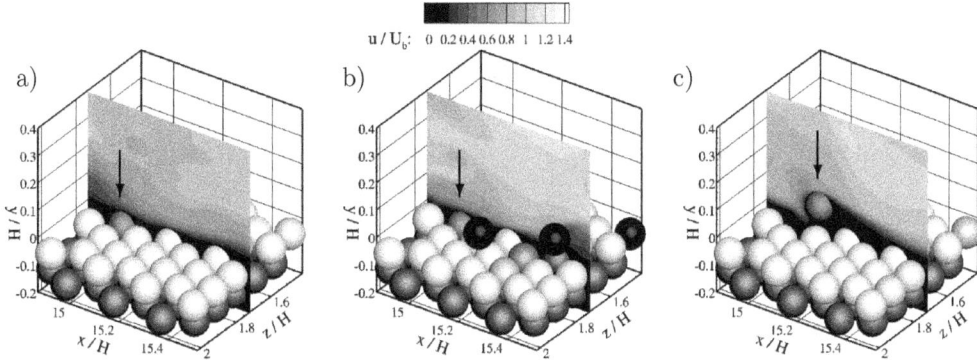

Figure 6.2: *Zoom into the domain shown in Fig. 5.2, displaying the single erosion event considered in Figs. 6.1 with the three marked instants in time A: $(t - t_{init}) = 200.5H/U_b$, B: $(t - t_{init}) = t_e = 203.2H/U_b$, and C: $(t - t_{init}) = 204.0H/U_b$. Particle colors: gray = fixed, white = $|u_p| < 1.5u_\tau$, black = $|u_p| > 1.5u_\tau$. The particle investigated is colored in red and indicated with an arrow.*

increase of these two quantities well above zero.

Here, the criterion to detect erosion events was chosen to be $y_p > 0$ and $u_p > u_\tau$. In the case shown in Figs. 6.1 and 6.2, the erosion event was triggered by a collision of an already eroded, mobile particle with the sediment bed in the vicinity of the particle in question, reflected by the spike in u_p/U_b around $(t - t_{init}) = 203.2H/U_b$ (instant B in Fig. 6.1). This lifts the particle off its initial position as shown by the curve of y_p. Once, the particle is slightly lifted above the resting position, it protrudes into the flow so that fluid forces onto the particle are substantially enhanced [55]. If the dislocation is large enough, the fluid forces can become sufficiently large to prevent the particle from falling back into its pocket, which indeed is observed in this case. The particle then starts to travel across the inactive bed with a saltating motion.

During the course of the simulation, continuous recording of the particle trajectories was performed and a total number of 340 of erosion events were detected. They constitute the samples of the present statistical analysis. Furthermore, continuous recording of fluid data, albeit coarsened by a factor of five for the sake of limited hard-disc storage, was carried out for the last $140H/U_b$, which allowed detailed investigation of the velocity fields for $n_{aver} = 157$ erosion events. This is a substantial improvement over experimental studies, such as, e.g., [124], who based their evaluations on a total of 26 erosion events for a dense packing. The present sample size allows for conditional averaging of quantities related to the disperse phase as well as to the continuous phase surrounding the particle being eroded. To perform statistical averaging of the detected events, a local coordinate system $x_{loc} = x - x_e$, $z_{loc} = z - z_e$, and $t_{loc} = t - t_e$ is defined with $x_e = x_p(t_e)$ and $z_e = z_p(t_e)$ the center point coordinates at the start of the erosion of the considered particle. No transformation was performed for the wall-normal coordinate y. The start of erosion at $t = t_e$ (instant B in Fig. 6.1) was designated as the instant of time with the first local minimum of the wall-normal position y_p of the particle when going backwards in time from the instant, when the criteria of erosion are met. The extent of the local averaging domain was chosen to be $-10D < x_{loc} < 10D$ and $-2.7D < z_{loc} < 2.7D$ in streamwise and spanwise direction, respectively, and covers the full channel height.

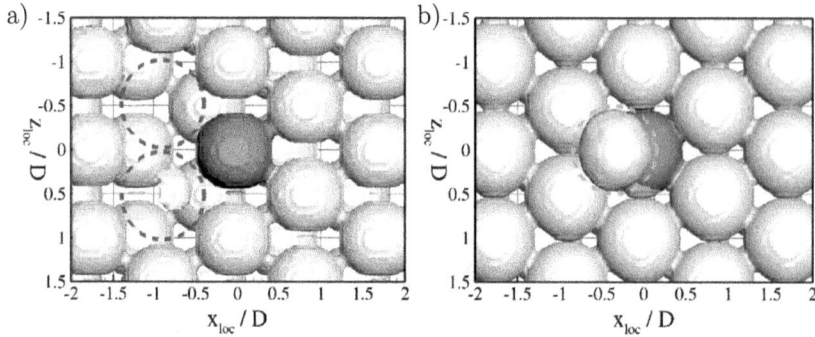

Figure 6.3: *Top view of the iso-surface of the ensemble-averaged porosity for two different thresh-olds: a) $\phi_T = 0.1$, and b) $\phi_T = 0.5$. The investigated particle is colored in red, the blue dashed circles indicate the gap in the upstream front of the particle, the green dashed circle indicates the colliding particle.*

The typical ensemble-averaged particle configuration of the closed bed is addressed by the porosity $\phi_T = n_f / n_{aver}$ with n_f the number of events, in which a grid cell is occupied by fluid. A close-up of the local averaging domain is shown in Fig. 6.3 illustrating the iso-surface of ϕ_T for two different thresholds in the vicinity of the eroded particle (colored in red). As expected, the porosity reflects the very regular pattern imposed by the hexagonal geometry of the fixed bed. Three principle types of events seem to be possible: erosion from the plane bed without involvement of another particle, collision with a mobile particle trig-gering the erosion event, and erosion in the presence of a gap, as analyzed with Fig. 6.3. A low threshold, indicating a high probability of a particle resting in a pocket of the hexagonal packing, reveals a gap in the upwind front of the eroded particle (blue circles in Fig. 6.3a). This means that, with some probability, the upstream positions are unoccupied resulting in an enhanced exposure of the particle to the mean flow, which increases the probability for mobilization [55]. If the threshold of the iso-surface is increased to $\phi_T > 0.2$, however, the gap is not present anymore (not shown here). Hence, the probability encountering this configuration during an erosion event is low. The higher threshold of $\phi_T > 0.5$ reveals that there is a high probability of finding a particle on top of the eroded one at the instant of erosion (Fig. 6.3b). It was indeed observed that in the present simulation 98 % of the erosion events were triggered by this mechanism. This particle is fully exposed to the mean flow and travels across the closed bed. It was reported in Fig. 5.14 in Chap. 5, however, that for this scenario the particle Reynolds number does not exceed a value of 120, so that according to [39], the fluid modification falls into a steady-wake regime with the extent of wakes of less than $1D$ in the direction. Hence, the impact of the colliding particle on the overall process is small, and it acts merely as a triggering feature but cannot by itself be the cause for the erosion. As discussed below, strong fluid forces exerted on the particle are necessary to push it out of its pocket. Nevertheless, moving particles, collide very frequently with the closed bed in the present regime, thus transferring kinetic energy to the particles resting in the closed bed. This feature increases the probability for entrainment as the collision loosens up the packing of the resting particles.

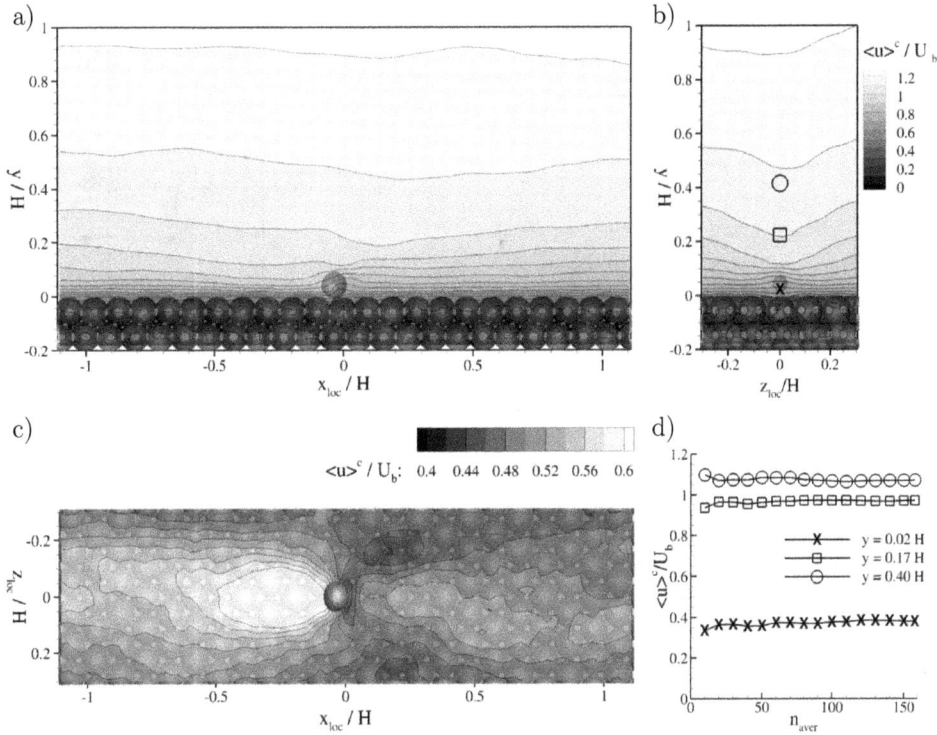

Figure 6.4: *Ensemble averaged fluid velocity* $\langle u \rangle^c$ *at an erosion event* ($t = t_e$) *and the iso-surface* $\phi_T = 0.5$ *of the porosity. a) center plane* $z_{loc} = 0$, *b) cross plane at* $x_{loc} = 0$, *c) top-view with contour at* $y = 0.023H$, *d) statistical convergence for the locations marked by the symbols in b).*

6.2.2   Ensemble averaged velocity fields

The classical approach to describe a criterion for incipient motion is based on mean quantities, such as the wall shear stress, describing the global average for an entire bed. It has been acknowledged in the literature, that these measures are insufficient to precisely predict particle entrainment [24, 18]. A stochastic approach, based on the probability density function focuses on extreme flow events within the turbulent fluctuations. This is not the complete picture as it has become common sense that extreme flow events of a sizable duration are responsible for erosion [48]. In this subsection, the conditionally averaged local field of the three velocity components at the instant of an erosion event are investigated to further explore quantities of the typical flow structures during an erosion event. In the following

$$\langle \theta \rangle^c = \frac{1}{n_f} \sum_{i=1}^{n_f} \theta(t_e) \tag{6.1}$$

denotes the conditional averaging to distinguish it from the averaging operators defined in Chaps. 4 and 5 if not stated otherwise.

The local field of ensemble averaged streamwise fluid velocity at the instant of an erosion event is illustrated in Fig. 6.4. The upstream flow at this instant in time is accelerated towards x_{loc}, while the fluid downstream of the particle is decelerated (Fig. 6.4 and c). The top-view in Fig. 6.4c illustrates immediately that the triggering particle shown in Fig. 6.3b is transported by a high speed streak typically being located in the up most downstream

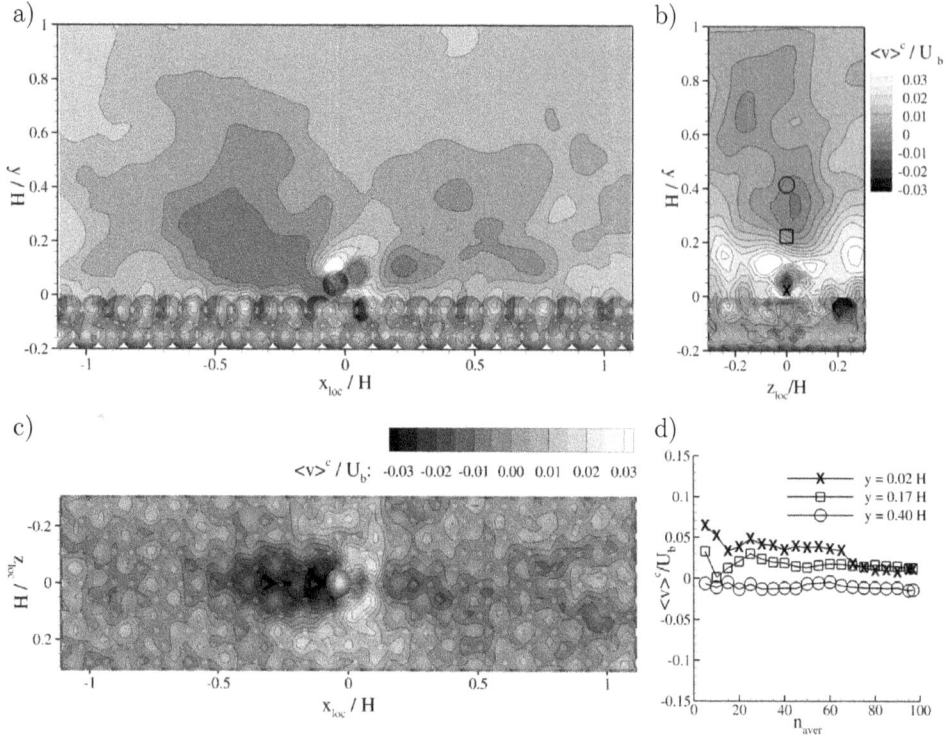

Figure 6.5: *Ensemble averaged fluid velocity $\langle v \rangle^c$ at an erosion event $(t = t_e)$ and the iso-surface $\phi_T = 0.5$ of the porosity. a) center plane $z_{loc} = 0$, b) cross plane at $x_{loc} = 0$, c) top-view with contour at $y = 0.023H$ (scales are adjusted for illustration purposes), d) statistical convergence for the locations marked by the symbols in b).*

front of that fluid structure. The high-speed streak has an streamwise extent larger than $1H$ in the upstream direction and shows a high degree of symmetry in the spanwise direction. It extends into the outer flow of the channel indicated by distinct spanwise gradients in the cross plane depicted in Fig. 6.4b. After the triggering particle has loosened up the packing, the high-speed fluid is subsequently going to flow across the eroded particle inducing high viscous stresses on the phase boundary. To investigate the degree of convergence of the present analysis, three locations of that plane were selected at $y = 0.02H$, $y = 0.17H$, and $y = 0.4H$ with the streamwise and spanwise coordinates chosen to be $x_{loc} = z_{loc} = 0$. This convergence study was carried out for all three velocity components, to illustrate the possible uncertainties that remain with the present statistical analysis. All three selected points show a high degree of convergence for the three locations and the 157 samples available.

The same conditional average was computed for the wall-normal component of the fluid velocity vector. This analysis indicates, that in addition to the high-speed streamwise velocity, a strong negative wall-normal velocity is reported in Fig. 6.5b for the same region just upstream of the eroded particle transporting the triggering particle. Hence, this region can be characterized as a strong sweep event [51, 133] that reaches far into the outer flow of the channel (Fig. 6.5a). This means, that fluid is going to flow into the pocket of the eroded particle after the triggering mechanism discussed above took place. In addition, two more regions become evident. An outward ejection occurs in the region $x_{loc} \approx 0$ that extends across the full spanwise extent of the local averaging domain, i.e. $[-0.3H; 0.3H]$. The

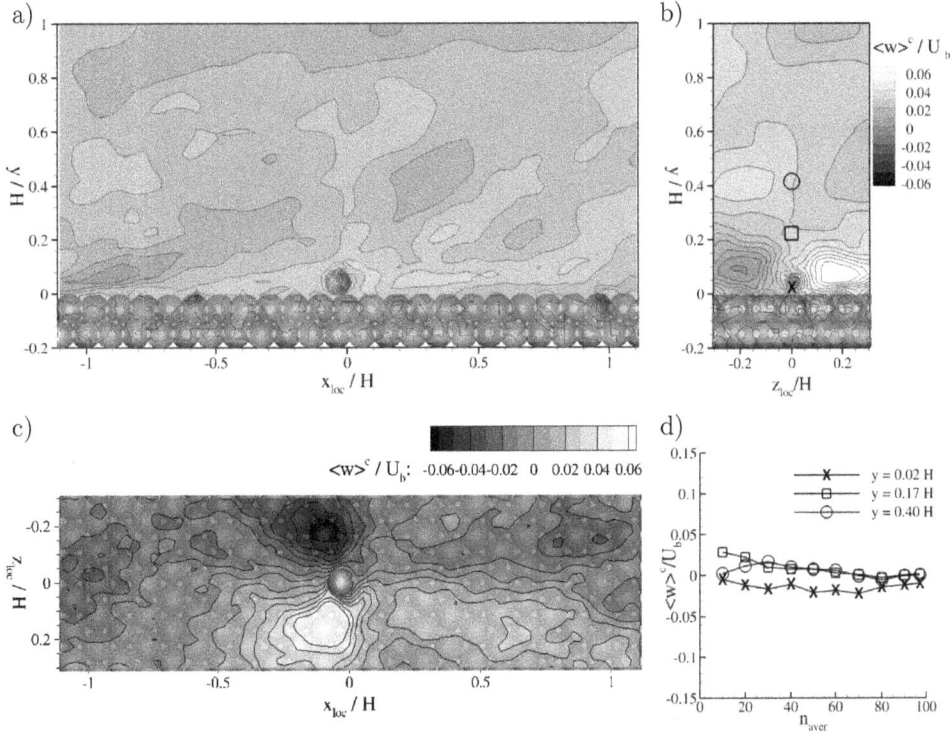

Figure 6.6: *Ensemble averaged fluid velocity $\langle w \rangle^c$ at an erosion event $(t = t_e)$ and the iso-surface $\phi_T = 0.5$ of the porosity. a) center plane $z_{loc} = 0$, b) cross plane at $x_{loc} = 0$, c) top-view with contour at $y = 0.023H$, d) statistical convergence for the locations marked by the symbols in b).*

large spanwise extent of this event proves that it is not only a local feature introduced in the vicinity of the triggering particle because the fluid has to flow around it. Instead, the triggering particle moves in a 'bursting' fluid event, i.e. an outward eruption of near-wall fluid. The second region, again, is a sweep event downstream of the particle with lower intensity than the one located upstream. Good convergence is reported for the outer flow (squares and circles in Fig. 6.5d) albeit the number of samples available was reduced for the wall-normal and spanwise component. For these components, only a total number of $n_{aver} = 97$ samples was available due to the sampling strategy that was modified during the course of the simulation. Although lower porosities, and hence less samples, are reported at the point in the near-wall region at $y = 0.02H$, a decent statistical convergence is reached for the given data set.

The ensemble averaged spanwise component of the fluid velocity vector shown in Fig. 6.6b and c reveals diverging fluid in the vicinity of the eroded particle. This was expected, because as soon as the sweep event discussed above hits a wall, energy transfer from the vertical component to the tangential ones must take place due to continuity of mass [110]. This even amplifies the effect of the loosened packing and illustrates that the spanwise coordinate of the core of the high-speed fluid structure must be close to the spanwise center coordinate of the eroded particle to have an effective attacking point. The degree of convergence is very high for all three components as indicated in Fig. 6.6d.

The results presented in Fig. 6.4–6.6 indicate a remarkable uniformity of the flow structures. Similar observations were made by Chan-Braun *et al.* [35], who investigated an open channel

a)

b)

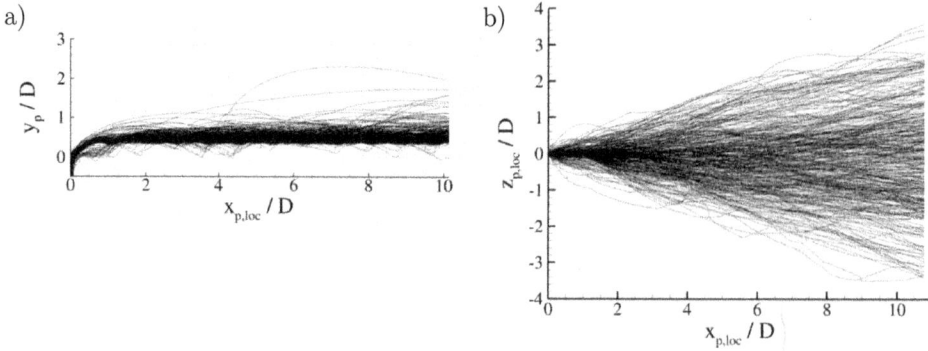

Figure 6.7: *Ensemble of the trajectories directly after mobilization. a) wall-normal position, b) spanwise position.*

flow over a fixed bed of square arrangement in the transitionally rough regime. In this study, strong drag and lift forces were used as an indication for possible erosion events. For such events, fluid structures elongated in streamwise direction were reported, which supports the present results. The discussed temporal and spatial scales are also in line with the findings of [45], who advocated for a fixed bed in the fully rough regime that high speed events with a duration of up to 3.5 bulk units cause pressure drops that may destabilize the sediment. While this reference draws a parallel with the so-called Bernoulli effect, i.e. a high velocity corresponds to low pressure, the present analysis for erosion events in the transitionally rough regime suggest a mechanism that is more differentiated. It was shown for the present dataset, that a succession of bursts and sweeps together with a triggering particle is needed for the data investigated, in order to obtain erosion events.

6.2.3 Particle diffusion

As discussed above, mobile particles play an important role in the present flow by triggering the erosion of resting particles, due to the transfer of kinetic energy by collision. It is therefore interesting to further investigate the particle trajectories. The full set of 340 samples is illustrated in Fig. 6.7 by means of the particle center coordinates. Only five trajectories that were substantially influenced by colliding with other traveling particles were excluded from the analysis.

Considering the evolution of the wall-normal coordinate y_p, two modes of transport can be identified after mobilization (Fig. 6.7a). First, particles with a wall-normal position of $y_p \leq 0.5D$ collide very frequently with resting particles exhibiting strong frictional forces due to collision. The overall motion, hence can be considered as rolling or sliding. Secondly, particles with $y_p > 0.5D$ have a large jumping length between two successive collisions and only few collisions per distance traveled [89]. This is characteristic for a saltating motion. For these particles, the maximum jumping height reaches up to $y_p = 2.4D$. This observation is in agreement with considerations of Nikora *et al.* [114], who found these two modes to be dominant for weak bed-load transport conditions. As expected, the first mode is more frequent than the second one in the present flow due to the low mobility of the particles. Looking into the spanwise direction, particle trajectories are fanning out with an even likelihood in positive and negative direction and a maximum displacement of $\pm 3.5D$

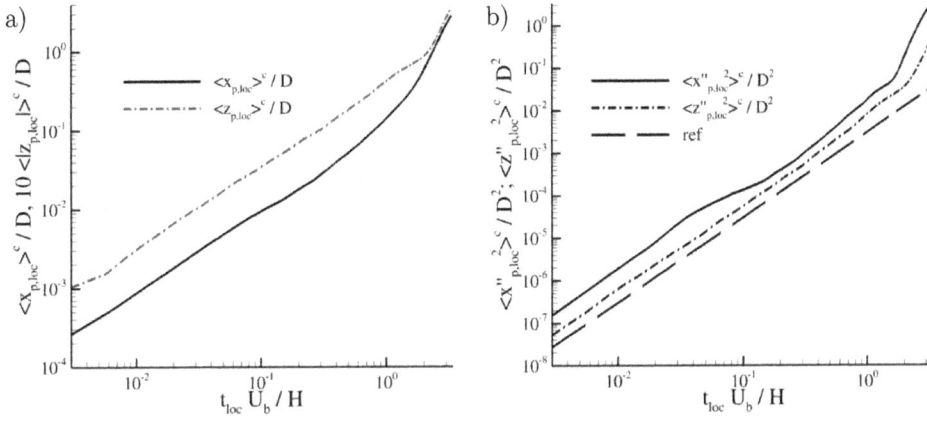

Figure 6.8: *Central statistics of the coordinates of entrained particle over time. a) Mean position (note the absolute value for spanwise position and the different scaling for streamwise and spanwise), b) Second order statistics. The curve "ref" corresponds to t^{n,γ_i} with $n = 2$ and $\gamma_i = 1$.*

over a distance of $11D$ in streamwise direction (Fig. 6.7b).

The presented ensemble of trajectories can be further examined in terms of their spread, termed diffusion in [113]. Diffusional movement can be interpreted as a Brownian motion, which is a passive propagation in time of particles submerged in a fluid [95]. This concept can be applied to the present study, where an ensemble of independently sampled particles of a finite size are transported by the turbulent channel flow. Statistically, the trajectories of the eroded particle can be well described by the central statistical moments of the quantity θ_p of that particle

$$\langle \theta_p''^n \rangle^c(t_{loc}) = \frac{1}{n_{aver}} \sum_{i=1}^{n_{aver}} (\theta_p(t_{loc}) - \langle \theta \rangle_p(t_{loc}))^n \qquad . \qquad (6.2a)$$

Here, n is the order ot the statistical moment and

$$\langle \theta \rangle_p^c(t_{loc}) = \frac{1}{n_{aver}} \sum_{i=1}^{n_{aver}} \theta_p(t_{loc}) \qquad\qquad (6.2b)$$

represents the first raw moment denoting the mean value at time t_{loc}. In particular, a diffusion process is characterized by the central moments of the local streamwise and spanwise particle position, i.e. $x_{p,loc}''^m$ and $z_{p,loc}''^m$, as they are expected to exhibit an increase in time obeying the proportionality $\langle x_{p,loc}''^m \rangle^c \propto t^{n\gamma_x(n)}$ and $\langle z_{p,loc}''^m \rangle^c \propto t^{n\gamma_z(n)}$ with $\gamma_x(n)$ and $\gamma_z(n)$ diffusion parameters for the streamwise and spanwise direction, respectively. These diffusion parameters may in principle be a function of the order of the statistical moment and differ in different directions. For isotropic conditions, however, the diffusion process is described by a single diffusion parameter γ_i, which yields $\gamma_i = \gamma_x = \gamma_z$.

For Brownian motion obeying Fick's law, one obtains $\gamma_x(n) = \gamma_z(n) = \gamma_i = 0.5$, indicating isotropic diffusion. According to [113], three more diffusion regimes can be identified for the case of particle transport. The first regime covers the trajectory between two successive collisions and is termed 'local' or 'ballistic' range. In this case, $\gamma_x = \gamma_z = \gamma_i = 1$ is obtained. The second regime describes a trajectory between two successive resting periods, while the third regime characterizes the full particle motion. Here, the diffusion parameter becomes $\gamma_i < 0.5$, because potential resting periods slow down the overall diffusion process.

The mean local streamwise particle position and the absolute value of the spanwise displacement over time are illustrated in Fig. 6.8a for the period of $0 < t_{loc} < 3.4\,H/U_b$. The mean local spanwise displacement exhibits a linear increase for the shown logarithmic scaling. The mean local streamwise particle position behaves in the same way until $t_{loc} = 0.4H/U_b$. For t_{loc} larger than this value the increase of $\langle x_{p,loc} \rangle^c$ grows with time. At $t_{loc} = 2.0H/U_b$, the diffusion process is enhanced even more as the spanwise displacement approaches the same increase as $\langle x_{p,loc} \rangle^c$. This effect can be attributed to the particles moving in a saltating state, as they speed up in time. The second order statistics follow the same proportionality of $t^{2\gamma}$ for both the streamwise and spanwise direction with $\gamma_x = \gamma_z = 1$ (Fig. 6.8b).

Hence, it can be deduced, that the presented results show isotropic scaling properties, i.e. $\gamma_x = \gamma_z = 1$ for an initial range. Moreover, the present ensemble of trajectories of incipient motion falls into the local, ballistic regime as the statistical quantities obey $\gamma_i = 1$ immediately after mobilization. During the course of the trajectories, saltating particles speed up more than rolling or sliding particles, so that particles propagate in anisotropic manner due to the different kinematic states observed.

6.3   Flow around dune-like clusters

6.3.1   Cluster tracking

In this section, the patterns observed for case *Ref* reported in Chap. 5 are further elucidated in terms of the characteristic flow features that possibly stabilize the cluster formation. As discussed above, the clusters observed for this case remind of two dune-like structures. These structures are constituted of mostly active particles such that the clusters propagate with a considerable velocity in streamwise direction across a layer of resting particles. Furthermore, it was found in Sec. 5.3.4 that these clusters have an average spacing of $12H = 0.5L_x$. For this physical scenario, no separation of the morphological and fluid-time scales occurs. Thus, the statistical tools presented in Chap. 4 would completely average out the typical pattern of the clusters. For an averaging interval sufficiently long, no particle clusters would be visible anymore, but the near-wall region would be described as a bed-load layer with homogeneous porosity in streamwise and spanwise direction. It was elucidated in detail in Chap. 5 by means of a Cartesian Distribution Function of the particles as well as the two-point correlation of the streamwise velocity component that indeed there are heterogeneous conditions in streamwise direction. The heterogeneity is introduced by the typical pattern formation described above and ultimately leads to a drastic increase in hydraulic resistance in terms of D^+ (cf. Tab. 5.4). This illustrates the strong impact of particle clustering on the overall behavior of the flow and raises the need to investigate the physical mechanisms that are responsible for the formation of the dune-like structures.

To maintain the typical pattern of the particle clusters within the averaging procedure, conditional averaging of the fluid information on a transformed coordinate system is needed. The transformed coordinates act as a frame moving with the characteristic velocity of the dune-like clusters to obtain conditions that are as stationary as possible. The moving frame transformation in turn requires a tool to track the center coordinate of the dune-like clusters.

For this purpose, a void fraction

$$\psi_y(x, z, t) = \frac{1}{H_{sed} + H} \int\limits_{-H_{sed}}^{H} (1 - \gamma(\mathbf{x}, t)) \, \mathcal{H}(|\mathbf{u}_p| - u_\tau) \mathrm{d}y \qquad (6.3)$$

integrated over wall-normal direction and conditioned by an indicator-function

$$\mathcal{H}(|\mathbf{u}_p| - u_\tau) = \begin{cases} 0, & \text{for } \mathbf{x} \text{ in particle } p \text{ and } |\mathbf{u}_p| < u_\tau \\ 1, & \text{for } \mathbf{x} \text{ in particle } p \text{ and } |\mathbf{u}_p| > u_\tau \end{cases} \qquad (6.4)$$

is defined. The indicator function $\mathcal{H}$ removes resting particles from the analysis and allows to investigate active particles only, as these were found to actually constitute the dune-like clusters. Subsequently, the spatial interval, in which the cluster is located, must be determined. As described above and reported in Sec. 5.3, nearly homogeneous conditions are reported for the spanwise direction (cf., e.g. the nearly wall-parallel iso-lines of porosity illustrated Fig. 5.15a). This simplifies the problem, since it can be assumed that the dune-like structures occupy the full spanwise extent, but have a finite size in streamwise direction. Hence, solely determination of the interval in streamwise direction (x_{start}, x_{end}) is required and the determination of the center of a particle cluster becomes a one-dimensional problem. Integrating $\psi(x, z, t)$ over this local interval yields

$$\Psi_i(z, t) = \int_{x_{start,i}}^{x_{end,i}} \psi(x, z, t) \mathrm{d}x \qquad . \qquad (6.5a)$$

The local center of mass of the particle cluster now simply becomes the center of mass as a function of the spanwise coordinate. An integral value can be obtained by averaging the center of mass over the spanwise extent, which yields

$$x_{core,i}(x, t) = \frac{1}{L_z} \int_0^{L_z} \frac{1}{\Psi(z, t)} \int_{x_{start,i}}^{x_{end,i}} \psi(x, z, t) \, x \, \mathrm{d}x \, \mathrm{d}z \qquad (6.5b)$$

for the center coordinate of a cluster with i the index of the cluster investigated. In what follows, the two clusters will be referred to as *dune 1* and *dune 2*. Note that these labels are meant as metaphors and the same precaution outlined in Sec. 5.3.1 applies to this analysis. The initial values of the tracking procedure were determined by the global maximum of the void fraction $\max\{\psi(x, z, t = t_{init} - 10H/U_b)\}$, which is 10 bulk units before the actual averaging time was started, to determine the streamwise coordinate of the first dune $x_{core,1}^{init}$. Furthermore, the initial coordinate of the second dune was set to $x_{core,2}^{init} = x_{core,1}^{init} + 0.5\,L_x$ according to the mean distance of $0.5L_x = 12H$ between the particle clusters. The initial start and ending value of the integral in x-direction were set to $x_{start,1}^{init} = x_{end,2}^{init} = x_{core,1}^{init} - 6H$ and $x_{end,1}^{init} = x_{start,2}^{init} = x_{core,1}^{init} + 6H$. The first ten bulk units before the averaging time started were discarded to minimize the influence of the assumptions made for the initial conditions on the tracking routine.

The values chosen for the initialization is based on two assumptions. First, only minor deviations of the streamwise spacing occurs during the averaging time and, second, the respective amount of particles constituting *dune 1* and *dune 2* is approximately equal. These assumptions, of course, are not going to hold for the course of the simulation, because exchange between the two large-scale clusters is possible changing the associated number of

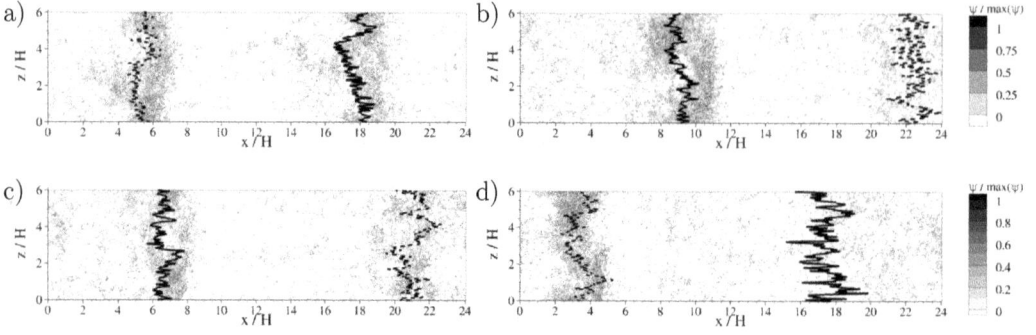

Figure 6.9: *Instantaneous distribution of the void fraction ψ normalized by its maximum value for four snapshots: a) $t - t_{init} = 149H/U_b$, b) $t - t_{init} = 441H/U_b$, c) $t - t_{init} = 545H/U_b$, and d) $t - t_{init} = 685H/U_b$. The solid and dashed lines indicate the center of mass of the first and the second structure, respectively.*

particles. Also deviations of the cluster spacing may occur. Therefore, a dynamic procedure for the determination of the interval (x_{start}, x_{end}) is required to account for these effects. The boundary values of the interval were adjusted by using the center-coordinates between the center of mass of the dune-like structures. Since these coordinates are not known *a priori*, the coordinates of the foregoing sample at $t - H/U_b$ were used, which yields

$$x_{start,1}(t) = \frac{1}{2}\left(x_{core,1}(t - H/U_b) + x_{core,2}(t - H/U_b) - (2n_{ftt} + 1)L_x\right) \tag{6.6a}$$

$$x_{end,1}(t) = \frac{1}{2}\left(x_{core,2}(t - H/U_b) + x_{core,1}(t - H/U_b) - (2n_{ftt})L_x\right) \tag{6.6b}$$

$$x_{start,2}(t) = x_{end,1}(t) \tag{6.6c}$$

$$x_{end,2}(t) = x_{start,1}(t) \tag{6.6d}$$

with the current spacing at time t being based on the cluster coordinates at $t - H/U_b$. Furthermore, n_{ftt} is the number of flow through times of a particle structure, which is the number a structure has traveled through the entire streamwise extent of the computational domain. With this definition, accessing information of a neighboring cluster is avoided and it allows a strict distinction between the two dune-like structures.

The time-scales of processes like the exchange of particles and the related growth and decay of large-scale clusters are known to be quite large [160, 9]. The simulation time was, therefore, extended from $260H/U_b$ to more than $680H/U_b$. Four characteristic examples of the distribution of ψ as well as the center of mass of particle clusters are illustrated for case *Ref* in Fig. 6.9. These four situations were selected progressing in time and they represent different states. First, an equilibrium between the two dunes was observed (Fig. 6.9a). This is reflected by a roughly equal distribution of the amount of particles constituting a cluster. This condition, however, eventually becomes unstable and the symmetry of the two clusters breaks up for $t - t_{init} > 260H/U_b$. This is illustrated in Fig. 6.9b, where *dune 1* has grown on the expense of *dune 2*. Due to the increased amount of particles constituting *dune 1*, the characteristic shape of the center of mass is smoothened. The second cluster, on the other hand, diminishes and only few particles can actually act as a roughness element. Later in time, this ordering breaks up again and two even structures emerge (Fig. 6.9c) that show a similar ordering observed for the situation in Fig. 6.9a. This configuration evolves even

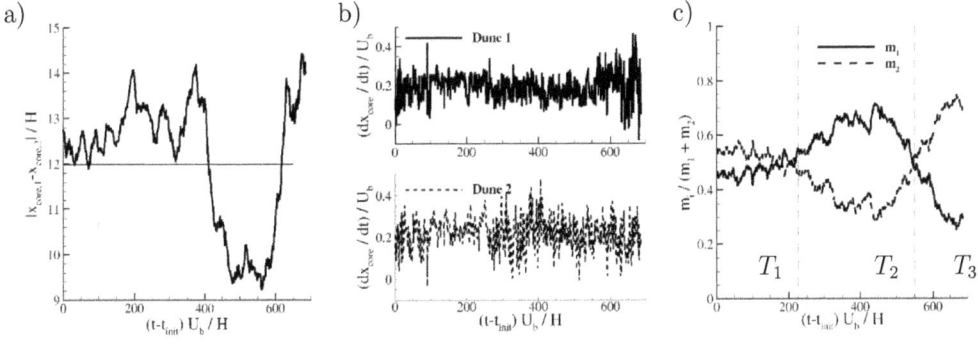

Figure 6.10: *Time evolution of the two major clusters observed in case* Ref: *a) distance between the clusters, b) cluster velocity, and c) cluster mass according to Eq. (6.7). The dashed vertical line indicates the starting and ending time of the three time intervals identified.*

further to a large *dune 2* and a small irregular *dune 1* (Fig. 6.9d). This situation is the inverse to Fig. 6.9b, where the first cluster is larger than the second.

The temporal evolution of *dune 1* and *dune 2* can be further investigated by the distance $|x_{core,1}(t) - x_{core,2}(t)|$ between the two clusters and the related cluster velocity. The exchange processes are addressed by the associated mass of a cluster

$$m_i(t) = \rho_p \int_0^{L_z} \int_{-H_{sed}}^{H} \int_{x_{start,i}}^{x_{end,i}} (1 - \gamma(\mathbf{x}, t)) \, \mathcal{H}(\mathbf{x}, t) \mathrm{d}x \, \mathrm{d}y \, \mathrm{d}z \qquad , \qquad (6.7)$$

which gives a measure to describe the mass fraction of a cluster as $m_i(t)/(m_1(t) + m_2(t))$. During the state of equilibrium (Fig. 6.9a), which lasts until $t - t_{init} \approx 230H/U_b$, the distance between *dune 1* and *dune 2* is fairly constant and the amount of particles per cluster is close to an equal distribution with a spacing close to $0.5L_x = 12H$ (denoted by the vertical line and T_1 in Fig. 6.10c). As soon as *dune 1* starts to grow, the distance between the cluster starts to increase as well. This interval is labeled T_2. Later in time at $t - t_{init} \approx 400H/U_b$, *dune 2* starts to gain mass from *dune 1* until another unstable state of equilibrium is reached at $t - t_{init} \approx 550H/U_b$. The remaining sampling time denoted as T_3 is dominated by a large *dune 2* and a small irregular *dune 1*. Hence, after a given initial phase of equilibrium, the present physical configuration undulates between the two extremes with one dune-like cluster containing a lot more particles than the other.

6.3.2   Conditional averaging and mean shape of particle cluster

As demonstrated by Fig. 6.9, the procedure outlined in Sec. 6.3.1 allows for determination of the center of mass of a particle cluster for a given instant in time. Moreover, characteristic velocities of the particle clusters can be assigned, which is the desired information to perform three-dimensional conditional averaging of fluid quantities in a coordinate system moving along with the mid-point between the two dunes. For this reason, a shift of the global coordinate system is performed via

$$\begin{aligned} x^c &= x + \frac{1}{2}(x_{core,1} + x_{core,2}) \qquad , \\ y^c &= y \qquad , \\ z^c &= z \qquad . \end{aligned} \qquad (6.8)$$

n	$t_{n,start}$	$t_{n,end}$	$T_n U_b/H$	$\bar{x}_{core,1}/U_b$	$\bar{x}_{core,2}/U_b$	$\overline{|x_{core,2}-x_{core,1}|}/H$
1	0	227	227	0.210	0.208	12.7
2	227	549	322	0.213	0.225	11.8
3	549	685	136	0.250	0.217	11.9

Table 6.1: *Intervals for time-averaging and the related velocities of the particle clusters and their related distance. Overbar denotes time-averaging over T_n.*

The superscript c indicates the conditioned streamwise coordinate. Note that this transformation is relevant for the streamwise coordinate only and in the following, the superscript will be omitted for the wall-normal and spanwise coordinates for brevity. With this transformation, the centers of *dune 1* and *dune 2* become approximately stationary in time with their location at $x^c \approx 18H$ and $x^c \approx 6H$, respectively, which fulfills the requirement of a moving frame of reference [64].

It was reported in Fig. 6.9 in a qualitative way and quantified by m_i in Fig. 6.10 that substantial exchange of mass between the two clusters occurs throughout the time simulated. This is a different situation compared to the pattern formation encountered in the simulations presented in Chap. 4. Therefore, it was decided to split the total averaging time $T_0 = T_{sample}$ into three consecutive time intervals with duration $T_n = t_{n,end} - t_{n,start}$ and $n \in \mathbb{N}$ the index increasing in time. This immediately requires that the duration $T_n(t) = t_{end}(t) - t_{start}(t)$ is not constant for any given n, but t_{start} and t_{end} must be chosen according to a suitable criterion to represent the different states discussed in Sec. 6.3.1 above:. i) a state of equilibrium, ii) *dune 1* being dominant over *dune 2*, and iii) *dune 2* being dominant over *dune 1*. It was shown in Fig. 6.10 that the ratio $m_i(t)/(m_1(t) + m_2(t))$ is a suitable criterion to describe these states. When this ratio becomes lower than 0.5 at $t - t_{init} \approx 227U_b/H$, an uneven particle distribution starts to develop. The particle distribution stays uneven until $t - t_{init} \approx 549U_b/H$. Note that this second state of equilibrium illustrated in Fig. 6.9c is not analyzed in the following, because this situation is not stable within a reasonable averaging interval. This yields the three time intervals T_1, T_2, and T_3 with the boundary values t_{start} and t_{end} of T_n as defined in Tab. 6.1. The time interval T_1 also corresponds to the averaging time used for the analysis of case *Ref* presented in Chap. 5 and reminds of the situation of a turbulent channel flow across streamwise periodic hills as investigated by [63, 22]. The time intervals T_2 and T_3 represent the situations of one cluster being dominant over the other as reflected by the associated masses m_i. The time-average of the cluster velocities of *dune 1* and *dune 2* shows that these situations go along with the condition of the dominant cluster slowing down, i.e. variations of the cluster spacing.

Analogous to Eq. (5.2) proposed in Sec. 5.3.2, the time-averaged porosity

$$\phi_T^c = \frac{1}{T_n} \int_{T_n} \gamma(x^c, y, z, t^*) \mathrm{d}t^* \quad , \tag{6.9}$$

gives the mean shape of a cluster within the n-th characteristic time interval T_n. The qualitative observations of Fig. 6.9 is now manifested in a statistical sense in Fig. 6.11 showing iso-surfaces of the porosity for a critical value of $\phi_T^c = 0.82$. As desired, dune-like clusters emerge on the locations at $x^c \approx 6H$ and $x^c \approx 18H$ introducing heterogeneity in streamwise direction. Their major axis, however, is oriented in spanwise direction so that roughly homogeneous conditions can be assumed in this direction. During T_1, the clusters

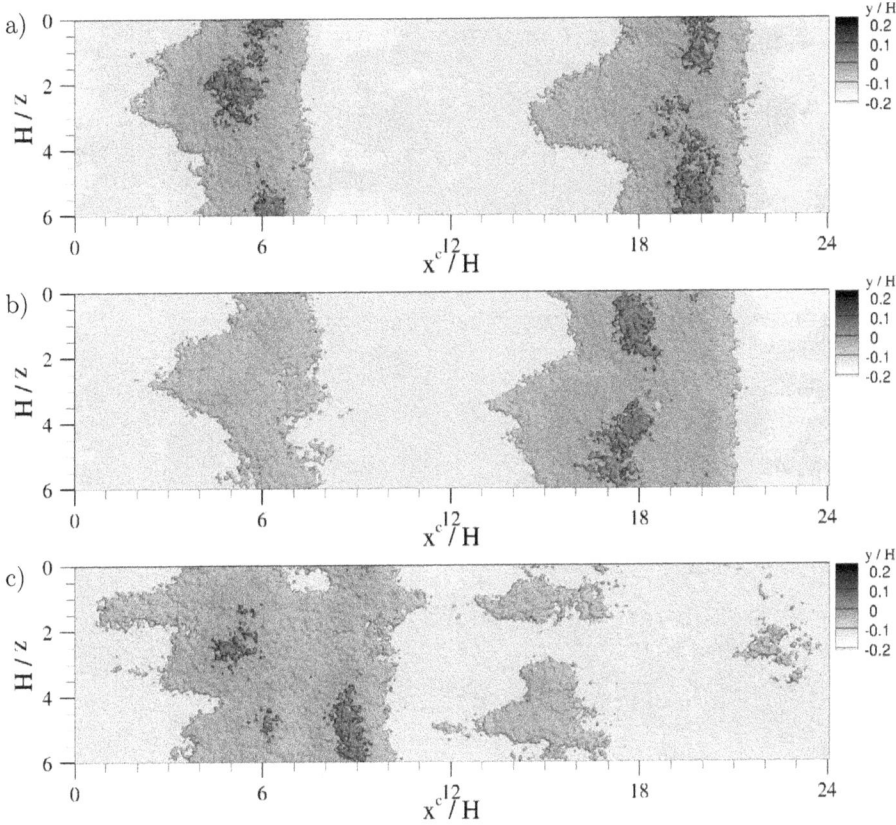

Figure 6.11: *Time-averaged shape of the particle structures observed illustrated by 3D-iso-surfaces of the porosity ($\phi_T^c = 0.82$) colored by the wall-normal elevation. a) T_1, b) T_2, and c) T_3.*

have a very similar shape with a wavy upstream front and very distinct downstream front. Between the time intervals T_2 and T_3, the dominant cluster grows in width, but not in height. The upstream front of the cluster remains somewhat fuzzy, while the downstream has a distinct edge. During T_3, *dune 1*, which ought to be located at $x^c \approx 18H$ disappeared completely, while *dune 2* has grown in width, but still shows a very patchy pattern with two regions, where the iso-surface of the porosity exceeds the height $y = 0.11H = 1D$, being located at $x^c \approx 5H$ and $x^c \approx 9H$. This may be due to the fact that time interval T_3 does not describe a full period, i.e. two instants of time with $m_i(t)/(m_1(t) + m_2(t)) = 0.5$ marking the boundary values of T_n (cf. the boundary values of T_2). This can be remedied with an extended time series, but could not be realized in the present study.

6.3.3   Time-averaged flow fields

The conditionally time-averaged flow fields can now be assessed by the same averaging procedure defined in Sec. 4.3.1. Applying the double-averaging operator to the present situation of conditioned coordinates and consecutive averaging of time intervals T_n yields

$$\langle \overline{\theta}^c \rangle^n = \frac{1}{V_{n,m}} \int_{V_0} \frac{1}{T_{n,f}} \int_{T_n} \theta(x_i^{c*}, t^*) \gamma(x_i^{c*}, t^*) \mathrm{d}t^* \mathrm{d}V^* \tag{6.10}$$

a)

b)

c)

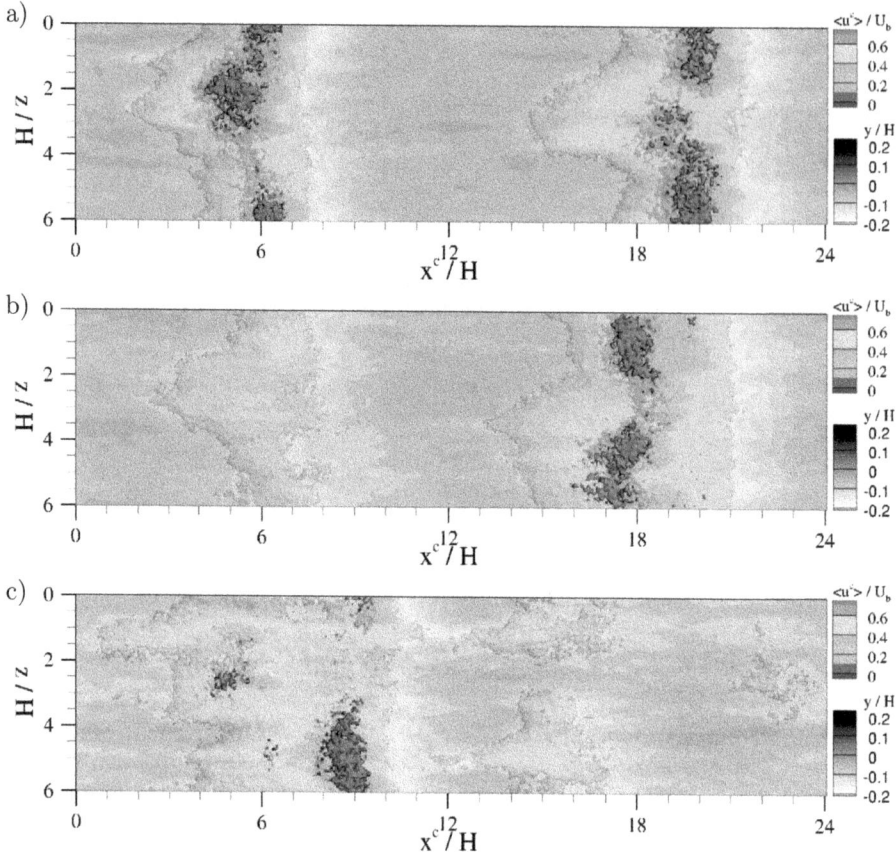

Figure 6.12: *Slice at $y = 1D$ of the double-averaged streamwise velocity for averaging volume $V_{0,3}$. a) T_1, b) T_2, and c) T_3. Time-averaged cluster shape as for Fig. 6.11.*

for the nth time interval. Recall that this procedure was referred to as intrinsic average, since only the part $V_{n,m}$ of the averaging volume V_0 that was occupied by fluid during T_n is accounted for. Two averaging volumes are considered. The volume $V_{0,3} = 0.5D \times \Delta_y \times 0.5D$ served as tool to convolute the highly-resolved three-dimensional data sets onto a coarsened three-dimensional grid, which is equivalent to apply some filtering to smoothen the signal. The second volume $V_{0,2} = 0.5D \times \Delta_y \times L_z$ is used to average across the spanwise direction, since it was shown in Sec. 6.3.2 that the roughness elements introduce heterogeneity in streamwise direction, but the flow remains fairly homogeneous in spanwise direction.

As expected, the particle clusters act as roughness elements that substantially slow down the surrounding fluid (Fig. 6.12) in the wall-normal level of $y = 1D = 0.11H$. This is especially true in the vicinity of the clusters. If the clusters shrink, their impact on the fluid decreases and vice-versa becomes stronger with increasing size. This effect, which is roughly homogeneous in spanwise direction, can be seen for all time intervals. In addition to these effects, the signal of the velocity shows streamwise streaks of low speed and high speed fluid during the interval T_3 shown in Fig. 6.12c. This suggests the development of secondary currents as already witnessed and discussed for Fig. 5.16 in Sec. 5.3. These low speed-streaks have the potential to locally slow down particle clusters, which leads to the mass exchanges between *dune 1* and *dune 2* reported in Fig. 6.10c. This observation is

a)

b)

c)

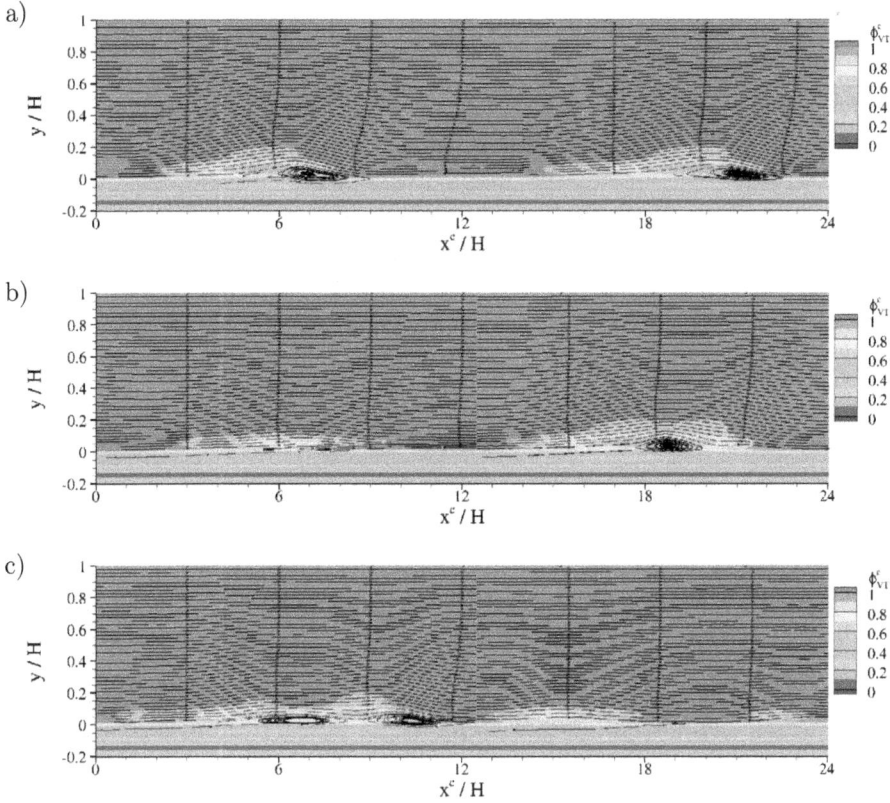

Figure 6.13: *Contour at $y = 1D$ of the double-averaged porosity for averaging volume $V_{0,2}$ and streamlines with streamwise u' and wall-normal v' velocity compoenents as defined by (6.11). a) T_1, b) T_2, and c) T_3.*

in line with the experiments reported in [81], who attributed these vortical structures the potential to entrain, deposit, and convey particles in the channel.

The impact of the clusters on the flow becomes even more evident, when looking at streamline plots of

$$u' = \left(\langle \overline{u}^c \rangle - \frac{1}{2} \left(\langle \overline{\tilde{x}}_{core,1} \rangle + \langle \overline{\tilde{x}}_{core,2} \rangle \right) \right) / U_b$$

$$v' = \langle \overline{v}^c \rangle / U_b \quad , \tag{6.11}$$

i.e. the averaged velocity of the particle clusters is subtracted from the signal of the streamwise component. These quantities were averaged over T_n and $V_{0,2}$ using the operator (6.10) yielding the illustration shown in Fig. 6.13.

In case of equilibrium (Fig. 6.13a), two recirculation zones of similar size are visible downstream of the clusters with a streamwise extent of roughly $4H$. Hence, the porous roughness elements decelerate the fluid that flows through them and cast a large wake in downstream direction. This pattern is similar to the fluid dynamics of stationary river dunes with a separation zone in the downstream region and a shear layer in the outer flow introduced by the roughness element [19]. Since in the present study it is shown that the particle clusters move in downstream direction, it is obvious, that small-scale particle clusters that enter the

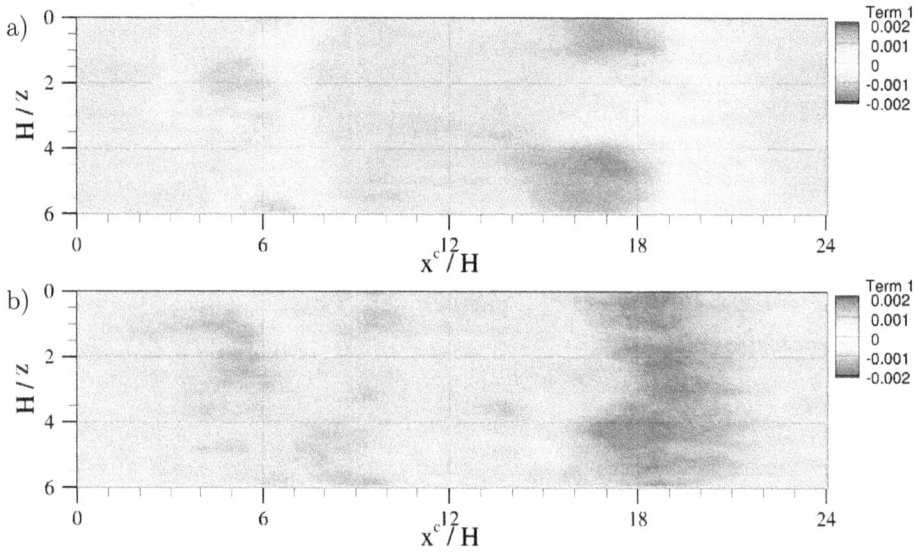

Figure 6.14: *Contour at $y = 1D$ of the local acceleration (term 1 of the double-averaged momentum balance (4.34)). a) $\Delta t = T_2 - T_1$, b) $\Delta t = T_3 - T_2$.*

separation zone are eventually caught up by the large scale clusters and subsequently merge with it. During T_2, the recirculation of *dune 2* decreased significantly in size, while the extent of the recirculation zone of *dune 1* was diminished and shifted towards the center of the cluster rather than in the downstream part as observed for T_1. This pattern is even more pronounced during T_3, where no recirculation is visible for *dune 1* and two weak recirculation zones become evident for *dune 2*. These recirculation zones are generated by two particle clusters that merge during T_3 due to their overlapping recirculation zones as illustrated by the massive single cluster shown in Fig. 6.9d.

Taking advantage of the long time signal of the present study and due to the unsteady behavior observed, it is now interesting to evaluate the first term of the double-averaged momentum balance (4.34), i.e. the local acceleration $\frac{\partial \phi_{Vm} \langle \phi_T \overline{u}_i \rangle}{\partial t}$. This gives a quantitative measure for the importance of this term to account for unsteady behavior of the flow. This was done by the approximation of a backward-difference scheme

$$\frac{\partial \phi_{Vm}^c \langle \phi_T^c \overline{u}_i^c \rangle}{\partial t} \approx \frac{\phi_{Vm}^{c,n} \langle \phi_T^{c,n} \overline{u}_i^{c,n} \rangle - \phi_{Vm}^{c,n-1} \langle \phi_T^{c,n-1} \overline{u}_i^{c,n-1} \rangle}{0.5(t_{n,end} - t_{n-1,start})} \qquad . \tag{6.12}$$

Fig. 6.14 shows the spatial distribution of the local acceleration in the near-wall region. The values are normalized by U_b^2/H to maintain non-dimensionality. Compared to the values reported Sec. 4.5.3, the local acceleration shows values that are one order of magnitude below the dominant terms of (4.34), which are the turbulent fluctuations and the viscous stresses. Nevertheless, for the given configuration, this term contributes a significant part to the momentum balance for the chosen discretization in time T_n and space $V_{0,3}$. As expected, strong negative values are reported in Fig. 6.14a. These are located in the area, where *dune 1* has grown in width. Since *dune 2* shrunk in size, accelerated fluid is reported in the region around $x^c = 6H$. The difference of the intervals T_2 and T_3 is shown in Fig. 6.14b. Here, a strong acceleration is reported for the region $x^c \approx 18H$, where *dune 1* was located during T_1 and T_2. In this area, the porosity has increased as well. The deceleration at $x^c \approx 6H$

is not as strong, because here decreased porosity balanced the decrease of $\langle \phi_f^c \overline{u}^c \rangle$. Hence, parts of the momentum, that were lost in this region must be taken up by the interfacial terms 10 and 11 of the double-averaged momentum balance, because the interfacial area S_{int} is directly proportional to the number of particles in this cluster. This illustrates the importance to account for the local acceleration, whenever one wants to describe unsteady flow conditions. It must be highlighted, however, that these changes happen on time-scales that exceed the simulation times reported in literature so far, e.g. [140, 91, 76, 155].

6.4   Conclusions

Two distinct mechanisms were investigated by means of conditional averaging analysis. Choosing appropriate conditions allowed for the investigation of the relevant mechanism in full detail. For erosion events of single particles out of a closed bed of resting particles, ensemble averaging was performed. It was found that most erosion events are triggered by collisions with active particles propagating with high velocities across the closed bed. This collision loosens up the close packing and must happen under certain flow conditions. At the instant of collision, a strong burst must happen at the location of the eroded particle followed by a strong sweep event that has a duration of several bulk units in order to have the potential to push the particle out of its pocket.

The flow mechanisms stabilizing dune-like particle clusters are addressed by a moving frame analysis yielding stationary locations of the large-scale particle clusters. The long term analysis revealed a mass exchange between the clusters. This is caused by low speed streaks that slow down small-scale particle clusters. As soon as these clusters enter the recirculation zone of the large scale dune-like structures they eventually are going to merge with it. It was demonstrated that this process is organized with a certain periodicity in time and a transient analysis becomes necessary to account for the significant local changes in time as is featured and demonstrated, e.g., in the DAM-framework.

7 Summary and outlook

The prediction of bed-load transport is a challenging task. The development of new approaches to describe the complex interaction between a turbulent flow and a multitude of particles requires data with a high resolution in time and space. This requirement is met by Direct Numerical Simulations. In this thesis, the Immersed Boundary Method is presented as a promising tool to provide phase-resolved data with full coupling of fluid-particle and particle-particle interaction. In this regard, collision modeling becomes a key issue to guarantee a sound four-way coupling of the flow. Methods like the simple Repulsive Force Model and the sophisticated Adaptive Collision Model are presented to achieve this and they were investigated for their impact on bed-load transport. Two test cases are considered: a trajectory of a single particle transported in a turbulent open-channel flow and the interaction of many mobile particles in a turbulent open-channel flow. Significant changes in the overall behavior as well as statistical quantities can be found for both scenarios depending on the choice of collision model. An improved physical realism was observed when the Adaptive Collision Model was used. Hence, one main conclusion of this thesis is that higher sophistication is needed for the modeling of particle-particle interaction when dealing with contact-dominated transport modes since significant

The enormous amount of data produced by the Direct Numerical Simulations need to be convoluted into a handy sets of quantities without loosing the general description of the flow behavior. The Double-Averaging-Methodology provides a sound platform based on physical reasoning rather than on empirical correlations and is applied for the first time in this thesis for the situation of a turbulent flow across a mobile granular bed. Using a variety of averaging strategies, a full description of the momentum balance is given together with the evaluation of the modified hydraulic resistance as well as the identification of large scale secondary currents. The Double-Averaging-Methodology has proven to be a very suitable framework to provide for the interpretation of the highly-resolved data.

Based on the observations for the test cases mentioned above, a systematic study of the key-parameters mobility and sediment supply for the scenario of bed-load transport in a turbulent open-channel flow is presented. This study employs a very large computational domain to allow particles to develop bed-forms with high fidelity. The pattern formation as well as the modification of fluid and particle related quantities are compared to the case of an unladen flow. This allows for the linkage of the modified length-scales of coherent fluid structures as well as particle clusters to the increased hydraulic resistance observed. These patterns are shown to be very stable in time and increase the bottom friction significantly.

To refine the analysis from the interpretation of the well-developed flow behavior to singular events that lead to local instabilities, tools for conditional averaging are presented. This facilitates the detection of erosion events and an analysis of the physical mechanisms leading to particle erosion on the one hand, and a detailed description of the development and evolution of large-scale particle clusters on the other. For the erosion events, a sequence of a strong burst and sweep event is responsible for particle erosion together with a colliding particle possibly triggering the event by a slight dislocation of the eroded particle. Large-scale clusters, in turn, are found to induce a separation zone in the downstream region that has the potential to assimilate small-scale particle cluster entering this region. This governs the mass-exchange between two successive clusters propagating in streamwise direction within periodic boundary conditions and yields an unstable equilibrium with a permanent growth and decay of the clusters.

Hence, the method of Direct Numerical Simulations of multiphase flows by using the Immersed Boundary Method together with the sophisticated Adaptive Collision Model, which was employed in the present thesis, has proven to be a valuable tool to provide the desired highly-resolved data of bed-load transport in turbulent open-channel flow in the transitionally rough regime. The high-fidelity data allows for comparison with experimental data gained in the fully rough regime. In addition, it opens up the possibility for further model development in the future. One example is the introduction of subgrid and supergrid models as suggested by [143, 145] in order to carry out simulations at high (realistic) Reynolds numbers. Another example is to describe the terms of the double-averaged momentum balance by means of advection-diffusion transport equations to provide a double-averaged model for engineers for flows of heterogeneous roughness expanding the classical RANS-simulations. The classical tool for engineers to predict the bed elevation of a movable bed, the Exner equation [123, 31] is of limited accuracy, as it presently suffers from uncertainties in quantifying the sediment exchange rates between the water column and the bed [150]. These exchange terms can be computed with the presented data sets in a straightforward manner and may serve as a benchmark to improve existing parameterizations. For future work to come, increasing computational capacity allows for simulations of another important key-parameter, i.e. the increased hydraulic roughness in terms of particle Reynolds number, on the formation of bed-forms [25]. Apart from the phenomenon of bed-load transport in open-channels, other applications of geo-physical flows, such as the models employed to describe turbidity currents [102] or granular flows [75] can benefit from the present study.

Bibliography

[1] http://acts.nersc.gov/hypre/, 2012.

[2] http://www.mcs.anl.gov/petsc/, 2012.

[3] http://tudresden.de/die_tu_dresden/zentrale_einrichtungen/zih, 2014.

[4] http://www.fz-juelich.de/ias/jsc/en, 2014.

[5] www.wikipedia.org/, 2014.

[6] R. J. Adrian. Hairpin vortex organization in wall turbulence. *Phys. Fluids*, 19(4):041301, 2007.

[7] R. J. Adrian, C. D. Meinhart, and C. D. Tomkins. Vortex organization in the outer region of the turbulent boundary layer. *J.Fluid Mech.*, 422(1):1–54, 2000.

[8] A. M. Ahmed and S. Elghobashi. Direct numerical simulation of particle dispersion in homogeneous turbulent shear flows. *Phys. Fluids*, 13(11):3346–3364, 2001.

[9] J. L. R. Allen. *Sedimentary structures: their character and physical basis*. Elsevier, 1982.

[10] M. Alletto and M. Breuer. One-way, two-way and four-way coupled LES predictions of a particle-laden turbulent flow at high mass loading downstream of a confined bluff body. *Int. J. Multiphase Flow*, 45:70–90, 2012.

[11] M. Amir, V. Nikora, and M. Witz. Saltation of sediment particles in turbulent open-channel flows: Effects of particle density and flow submergence. *J. Hydraul. Res.*, 2014, accepted.

[12] L. O. Amoudry. Assessing sediment stress closures in two-phase sheet flow models. *Adv. Water Resour.*, 48:92–101, 2012.

[13] C. Ancey and J. Heymann. A microstructural approach to bed-load transport: mean baviour and fluctutations of particle transport rates. *J. Fluid Mech.*, 744:129–168, 2014.

[14] R. A Bagnold. The flow of cohesionless grains in fluids. *Philos. T. Roy. Soc. A*, pages 235–297, 1956.

[15] R. A. Bagnold. An approach to the sediment transport problem from general physics. *US Geological Survey Professional Paper*, 422:231–291, 1966.

[16] R. A. Bagnold. The nature of saltation and of'bed-load'transport in water. *P. Roy. Soc. Lond. A Mat.*, 332(1591):473–504, 1973.

[17] S. Balachandar and J. K. Eaton. Turbulent dispersed multiphase flow. *Annu. Rev. Fluid Mech.*, 42:111–133, 2010.

[18] J.C. Bathurst. Effect of coarse surface layer on bed-load transport. *J. Hydraul. Eng.-ASCE*, 133:1192–1205, 2007.

[19] J. Best. The fluid dynamics of river dunes: A review and some future research directions. *J. Geophys. Res.: Earth Surfarce*, 110:F4, 2005.

[20] R. F. Blackwelder and R. E. Kaplan. On the wall structure of the turbulent boundary layer. *J. Fluid Mech.*, 76(01):89–112, 1976.

[21] P. Bradshaw. Turbulent secondary flows. *Annu. Rev. Fluid Mech.*, 19(1):53–74, 1987.

[22] M. Breuer, N. Peller, C. Rapp, and M. Manhart. Flow over periodic hillsânumerical and experimental study in a wide range of reynolds numbers. *Comput. Fluids*, 38(2):433–457, 2009.

[23] E. Buckingham. On physically similar systems; illustrations of the use of dimensional equations. *Phys. Rev.*, 4(4):345–376, 1914.

[24] J. M. Buffington and D. R. Montgomery. A systematic analysis of eight decades of incipient motion studies, with special reference to gravel-bedded rivers. *Wat. Resour. Res.*, 33(8):1993–2029, 1997.

[25] J. M. Buffington and D. R. Montgomery. Effects of hydraulic roughness on surface textures of gravel-bed rivers. *Wat. Resour. Res.*, 35(11):3507–3521, 1999.

[26] J. M. Buffington and D. R. Montgomery. Effects of sediment supply on surface textures of gravel-bed rivers. *Wat. Resour. Res.*, 35(11):3523–3530, 1999.

[27] B. Bunner and G. Tryggvason. Effect of bubble deformation on the properties of bubbly flows. *J. Fluid Mech.*, 495:77–118, 2003.

[28] S. M. Cameron, S. E. Coleman, B. W. Mellville, and V. I. Nikora. Marbles in oil, just like a river? In R. Ferreira, E. Alves, and A. Leal, J. Cardoso, editors, *River Flow 2006*, pages 927–935. Taylor & Francis, 2006.

[29] S. M. Cameron, V. I. Nikora, and S. E. Coleman. Double-averaged velocity and stress distributions for hydraulically-smooth and transitionally-rough turbulent flows. *Acta Geophysica*, 56(3):642–653, 2008.

[30] J. Campagnol, A. Radice, R. Nokes, V. Bulankina, A. Lescova, and F. Ballio. Lagrangian analysis of bed-load sediment motion: database contribution. *J. Hydraul. Res.*, 51(5):589–596, 2013.

[31] Z. X. Cao, G. Pender, and P. Carling. Shallow water hydrodynamic models for hyperconcentrated sediment-laden floods over erodible bed. *Adv. Water Resour.*, 29(4):546–557, 2006.

[32] A. O. Celik, P. Diplas, and C. L. Dancey. Instantaneous turbulent forces and impulse on a rough bed: Implications for initiation of bed material movement. *Wat. Resour. Res.*, 49(4):2213–2227, 2013.

[33] A. O. Celik, P. Diplas, C. L. Dancey, and M. Valyrakis. Impulse and particle dislodgement under turbulent flow conditions. *Phys. Fluids*, 22(4):046601, 2010.

[34] C. Chan-Braun, M. García-Villalba, and M. Uhlmann. Force and torque acting on particles in a transitionally rough open-channel flow. *J. Fluid Mech.*, 684:441–474, 2011.

[35] C. Chan-Braun, M. García-Villalba, and M. Uhlmann. Spatial and temporal scales of force and torque acting on wall-mounted spherical particles in open channel flow. *Phys. Fluids*, 25(7):075103, 2013.

[36] C. Chan-Braun, M. García-Villalba, and M. Uhlmann. Numerical simulation of fully resolved particles in rough-wall turbulent open channel flow. In *7th Int. Conf. Multiphase Flow*, Tampa, FL, 2010.

[37] F. Charru, B. Andreotti, and P. Claudin. Sand ripples and dunes. *Annu. Rev. Fluid Mech.*, 45:469–493, 2013.

[38] F. Charru, H. Mouilleron, and O. Eiff. Erosion and deposition of particles on a bed sheared by a viscous flow. *J. Fluid Mech.*, 519:55–80, 2004.

[39] R. Clift, J.R. Grace, and M.E. Weber. *Bubbles, Drops, and Particles*. Dover Publications, 1978.

[40] N. L. Coleman. A theoretical and experimental study of drag and lift forces acting on a sphere resting on a hypothetical streambed. In R. Ferreira, E. Alves, and A. Leal, J. Cardoso, editors, *Proc. 12th Congress of International Association for Hydraulic Research*, pages 185–192, 1967.

[41] S. E. Coleman, V. I. Nikora, S. R. McLean, and E. Schlicke. Spatially averaged turbulent flow over square ribs. *J. Eng. Mech.-ASCE*, 133(2):194–204, 2007.

[42] R.G. Cox and H. Brenner. The slow motion of a sphere through a viscous fluid towards a plane surface. Small gap widths, including inertial effects. *Chem. Eng. Sci.*, 22:1753–1777, 1967.

[43] C. Crowe. *Multiphase Flow Handbook*. CRC Press, 2006.

[44] J. J. Derksen. Simulations of granular bed erosion due to laminar shear flow near the critical shields number. *Phys. Fluids*, 23(11):113303, 2011.

[45] M. Detert, V. Nikora, and G. H. Jirka. Synoptic velocity and pressure fields at the water-sediment interface of streambeds. *J. Fluid Mech.*, 660:55–86, 2013.

[46] S. Dey, R. Das, R. Gaudio, and S. K. Bose. Turbulence in mobile-bed streams. *Acta Geophysica*, 60(6):1547–1588, 2012.

[47] W. E. Dietrich, J. W. Kirchner, H. Ikeda, and F. Iseya. Sediment supply and the development of the coarse surface-layer in gravel-bedded rivers. *Nature*, 340(6230):215–217, 1989.

[48] P. Diplas, C. L. Dancey, A. O. Celik, M. Valyrakis, K. Greer, and T. Akar. The role of impulse on the initiation of particle movement under turbulent flow conditions. *Science*, 322(5902):717–720, 2008.

[49] O. Durán, B. Andreotti, and P. Claudin. Numerical simulation of turbulent sediment transport, from bed load to saltation. *Phys. Fluids*, 24(10):103306, 2012.

[50] A. Dwivedi, B. W. Melville, A. Y. Shamseldin, and T. K. Guha. Analysis of hydrodynamic lift on a bed sediment particle. *J. Geophys. Res.: Earth Surface*, 116:F2, 2011.

[51] A. Dwivedi, B. W. Melville, A. Y. Shamseldin, and T. K. Guha. Flow structures and hydrodynamic force during sediment entrainment. *Wat. Resour. Res.*, 47(1):W01509, 2011.

[52] H.A. Einstein. The bed-load function for sediment transportation in open channel flows. *U.S. Dept. Agriculture, Soil Conservation Service Tech. Bull., 1026*, 1950.

[53] R. Ettema. *Hydraulic Modeling: Concepts and Practice*. American Society of Civil Engineers, 2000.

[54] E. A. Fadlun, R. Verzicco, P. Orlandi, and J. Mohd-Yusof. Combined immersed-boundary finite-difference methods for three-dimensional complex flow simulations. *J. Comput. Phys.*, 161(1):35–60, 2000.

[55] J. D. Fenton and J. E. Abbott. Initial movement of grains on a stream bed - effect of relative protrusion. *Proceedings of the Royal Society of London Series A-mathematical Physical and Engineering Sciences*, 352(1671):523–537, 1977.

[56] J.H. Ferziger and M. Perić. *Computational Methods for Fluid Dynamics*. Springer Verlag, 2002.

[57] J. Finnigan. Turbulence in plant canopies. *Annu. Rev. Fluid Mech.*, 32(1):519–571, 2000.

[58] J. J. Finnigan, R. H. Shaw, and E. G. Patton. Turbulence structure above a vegetation canopy. *J. Fluid Mech.*, 637:387–424, 2009.

[59] Y. Forterre and O. Pouliquen. Flows of dense granular media. *Annu. Rev. Fluid Mech.*, 40:1–24, 2008.

[60] P. Frey and M. Church. How river beds move. *Science*, 325:1509–1510, 2009.

[61] P. Frey and M. Church. Bedload: a granular phenomenon. *Earth Surf. Proc. Land.*, 63:58–69, 2011.

[62] J. Fröhlich. *Large Eddy Simulation turbulenter Strömungen*. Teubner Verlag, 2006 (in German).

[63] J. Fröhlich, C. P. Mellen, W. Rodi, L. Temmerman, and M. A. Leschziner. Highly resolved large-eddy simulation of separated flow in a channel with streamwise periodic constrictions. *J. Fluid Mech.*, 526:19–66, 2005.

[64] T. Gal-Chen. Errors in fixed and moving frame of references: Applications for conventional and doppler radar analysis. *J. Atmos. Sci.*, 39(10):2279–2300, 1982.

[65] M. Garcia. Sediment transport and morphodynamics. In M.H. Garcia, editor, *American Society of Civil Engineers, Manuals and Reports on Engineering Practice 110*, pages 21–168, 2008.

[66] M. Garcia-Villalba, A. G. Kidanemariam, and M Uhlmann. DNS of vertical plane channel flow with finite-size particles: Voronoi analysis, acceleration statistics and particle-conditioned averaging. *Int. J. Multiphase Flow*, 46:54–74, 2012.

[67] L. A. Giménez-Curto and M. A. C. Lera. Oscillating turbulent flow over very rough surfaces. *J. Geophys. Res.*, 101(C9):20745–20758, 1996.

[68] R. Glowinski, T. W. Pan, T. I. Hesla, D. D. Joseph, and J. Priaux. A fictitious domain approach to the direct numerical simulation of incompressible viscous flow past moving rigid bodies: Application to particulate flow. *J. Comput. Physics*, 169:363–426, 2001.

[69] R. Glowinski, T.-W. Pan, T.I. Hesla, and D.D. Joseph. A distributed Lagrange multiplier/fictitious domain method for particulate flows. *Int. J. Multiphase Flow*, 25:755–794, 1999.

[70] W. Gray and P. C. Y. Lee. On the theorems for local volume averaging of multiphase systems. *Int. J. Multiphase Flow*, 3(4):333–340, 1977.

[71] P. Gualtieri, F. Picano, and C. M. Casciola. Anisotropic clustering of inertial particles in homogeneous shear flow. *J. Fluid Mech.*, 629:25–39, 2009.

[72] H. Hertz. Über die Berührung fester elastischer Körper. *J. reine angew. Math.*, 92:156–171, 1882.

[73] S. Hoyas and J. Jiménez. Scaling of the velocity fluctuations in turbulent channels up to $Re_\tau = 2003$. *Phys. Fluids*, 011702, 2006.

[74] R. Jain. Erosion of single particles in a turbulent open channel flow. Master's thesis, Inst. f. Strömungsmech, TU Dresden, Germany, 2014.

[75] J. T. Jenkins and D. Berzi. Dense inclined flows of inelastic spheres: tests of an extension of kinetic theory. *Granul. Matter*, 12(2):151–158, 2010.

[76] C. Ji, A. Munjiza, E. Avital, J. Ma, and J. J. R. Williams. Direct numerical simulation of sediment entrainment in turbulent channel flow. *Phys. Fluids*, 25(5):056601, 2013.

[77] J. Jiménez. Turbulent flows over rough walls. *Annu. Rev. Fluid Mech.*, 36:173–196, 2004.

[78] J. Jiménez and P. Moin. The minimal flow unit in near-wall turbulence. *J. Fluid Mech.*, 225:213–240, 1991.

[79] G. Joseph and M. L. Hunt. Oblique particle wall collisions in a liquid. *J. Fluid Mech.*, 510:71–93, 2004.

[80] G. G. Joseph, R. Zenit, M. L. Hunt, and A. M. Rosenwinkel. Particle-wall collisions in a viscous fluid. *J. Fluid Mech.*, 433:329–346, 2001.

[81] D. Kaftori, G. Hetsroni, and S. Banerjee. Particle behavior in the turbulent boundary layer. i. motion, deposition, and entrainment. *Phys. Fluids*, 7(5):1095–1106, 1995.

[82] T. Kajishima and S. Takiguchi. Interaction between particle clusters and particle-induced turbulence. *Int. J. Heat Fluid Fl.*, 23(5):639–646, 2002.

[83] T. Kajishima, S. Takiguchi, H. Hamasaki, and Y. Miyake. Turbulence structure of particle-laden flow in a vertical plane channel due to vortex shedding. *JSME Int. J. B-Fluid T.*, 44(4):526–535, 2001.

[84] I. Karcz. Secondary currents and configuration of a natural stream bed. *J. Geophys. Res.*, 71(12):3109–3113, 1966.

[85] H. Kawamura, H. Abe, and Y. Matsuo. DNS of turbulent heat transfer in channel flow with respect to reynolds and prandtl number effects. *Int. J. Heat Fluid Fl.*, 20(3):196–207, 1999.

[86] T. Kempe. *A numerical method for interface-resolving simulations of particle-laden flows with collisions*. Dissertation, Technische Universität Dresden, 2011.

[87] T. Kempe and J. Fröhlich. Collision modelling for the interface-resolved simulation of spherical particles in viscous fluids. *J. Fluid Mech.*, 709:445–489, 2012.

[88] T. Kempe and J. Fröhlich. An improved immersed boundary method with direct forcing for the simulation of particle laden flows. *J. Comput. Phys.*, 231(9):3663–3684, 2012.

[89] T. Kempe, B. Vowinckel, and J. Fröhlich. On the relevance of collision modeling for interface-resolving simulations of sediment transport in open channel flow. *Int. J. Multiphase Flow*, 58:214–235, 2014.

[90] H. R. Khakpour, L. Shen, and D. K. P. Yue. Transport of passive scalar in turbulent shear flow under a clean or surfactant-contaminated free surface. *J. Fluid Mech.*, 670:527–557, 2011.

[91] A. G. Kidanemariam, C. Chan-Braun, T. Doychev, and M. Uhlmann. Direct numerical simulation of horizontal open channel flow with finite-size, heavy particles at low solid volume fraction. *New J. Phys.*, 15:025031, 2013.

[92] K. T. Kiger and C. Pan. Suspension and turbulence modification effects of solid particulates on a horizontal turbulent channel flow. *J. Turbulence,*, 3(19):1–17, 2002.

[93] J. Kim, P. Moin, and R. Moser. Turbulence statistics in fully developed channel flow at low Reynolds number. *J. Fluid Mech.*, 177:133–166, 1987.

[94] H.C. Kuhlmann. *Strömungsmechanik*. Pearson Studium, 2007.

[95] P.K. Kundu and I.M. Cohen. *Fluid mechanics*. Elsevier, 2008.

[96] E. Lajeunesse, L. Malverti, and F. Charru. Bed load transport in turbulent flow at the grain scale: Experiments and modeling. *J. Geophys. Res.: Earth Surfarce*, 115, 2010.

[97] P. Leopardi. A partition of the unit sphere into regions of equal area and small diameter. *Electron. Trans. Numer. Anal.*, 25:309–327, 2006.

[98] D. M. Lu and G. Hetsroni. Direct numerical simulation of a turbulent open channel flow with passive heat transfer. *Int. J. Heat Mass Transfer*, 38:3241 – 3251, 1995.

[99] C. Manes, D. Pokrajac, and I. McEwan. Double-averaged open-channel flows with small relative submergence. *J. Hyraul. Eng.-ASCE*, 133(8):896–904, 2007.

[100] C. Marchioli and A. Soldati. Mechanisms for particle transfer and segregation in a turbulent boundary layer. *J. Fluid Mech.*, 468:283–315, 2002.

[101] I. McEwan and J. Heald. Discrete particle modeling of entrainment from flat uniformly sized sediment beds. *J. Hyraul. Eng.-ASCE*, 127(7):588–597, 2001.

[102] E. Meiburg and B. Kneller. Turbidity currents and their deposits. *Annu. Rev. Fluid Mech.*, 42:135–156, 2013.

[103] E. Meyer-Peter and R. Müller. Formulas of bed-load transport. *Proceedings of the 2nd Meeting of the International Association for Hydraulic Structures Research*, pages 33–64, 1948.

[104] E. Mignot, E. Barthelemy, and D. Hurther. Double-averaging analysis and local flow characterization of near-bed turbulence in gravel-bed channel flows. *J. Fluid Mech.*, 618:279–279, 2009.

[105] J. Mohd-Yusof. Combined immersed boundary /B-Spline method for simulations of flows in complex geometries. *Center for Turbulence Research. Annual Research Briefs. NASA Ames / Stanford University*, pages 317–327, 1997.

[106] P. A. Moreno and F. A. Bombardelli. 3D numerical simulation of particle-particle collisions in saltation mode near stream beds. *Acta Geophysica*, 60(6):1661–1688, 2012.

[107] R. D. Moser, J. Kim, and N. N. Mansour. Direct numerical simulation of turbulent channel flow up to $Re_\tau = 590$. *Phys. Fluids*, 11:943–945, 1999.

[108] H. Mouilleron, F. Charru, and O. Eiff. Inside the moving layer of a sheared granular bed. *J. Fluid Mech.*, 628:229–239, 2009.

[109] M. Muste, K. Yu, I. Fujita, and R. Ettema. Two-phase flow insight into open-channel flows with suspended particles of different densities. *Environ. Fluid Mech.*, 9:161–186, 2009.

[110] I. Nezu and H. Nakagawa. *Turbulence in Open-Channel Flows*. IAHR/AIRH Monograph, 1993.

[111] N.Q. Nguyen and A.J.C. Ladd. Lubrication corrections for lattice-Boltzmann simulations of particle suspensions. *Phys. Rev. E*, 66:046708, 2002.

[112] V. Nikora, F. Ballio, S. Coleman, and D. Prokrajac. Spatially-averaged flows over mobile rough beds: definitions, averaging theorems, and conservation equations. *J. Hyraul. Eng.-ASCE*, 139(8):803–811, 2013.

[113] V. Nikora, H. Habersack, T. Huber, and I. McEwan. On bed particle diffusion in gravel bed flows under weak bed load transport. *Water Resour. Res.*, 38(6):1–17, 2002.

[114] V. Nikora, J. Heald, D. Goring, and I. McEwan. Diffusion of saltating particles in uni-directional water flow over a rough granular bed. *J. Phys. A- Math. Gen.*, 34(50):L743, 2001.

[115] V. Nikora, I. McEwan, S. McLean, S. Coleman, D. Pokrajac, and R. Walters. Double-averaging concept for rough-bed open-channel and overland flows: Applications. *J. Hyraul. Eng.-ASCE*, 133(8):884–895, 2007.

[116] V. Nikora, I. McEwan, S. McLean, S. Coleman, D. Pokrajac, and R. Walters. Double-averaging concept for rough-bed open-channel and overland flows: Theoretical background. *J. Hyraul. Eng.-ASCE*, 133(8):873–883, 2007.

[117] V. I. Nikora and P. M. Rowiński. Rough-bed flows in geophysical, environmental, and engineering systems: Double-averaging approach and its applications. *Acta Geophysica*, 56(3):529–533, 2008.

[118] Y. Nino and M. Garcia. Gravel saltation: 1. experiments. *Wat. Resour. Res.*, 30(6):1907–1914, 1994.

[119] Y. Nino and M. Garcia. Using lagrangian particle saltation observations for bedload sediment transport modelling. *Hydrol. Processes*, 12:1197–1218, 1998.

[120] Y. Nino and M. H. Garcia. Experiments on particle-turbulence interactions in the near-wall region of an open channel flow: implications for sediment transport. *J. Fluid Mech.*, 326:285–319, 1996.

[121] F. Osanloo, M. R. Kolahchi, S. McNamara, and H. J. Herrmann. Sediment transport in the saltation regime. *Phys. Rev. E*, 78:011301, 2008.

[122] M. Ouriemi, P. Aussillous, and E. Guazzelli. Sediment dynamics. part 1. bed-load transport by laminar shearing flows. *J. Fluid Mech.*, 636:295–319, 2009.

[123] C. Paola and V. R. Voller. A generalized exner equation for sediment mass balance. *J. Geophys. Res.*, 110(F4):F04014, 2005.

[124] A. N. Papanicolaou, P. Diplas, N. Evaggelopoulos, and S. Fotopoulos. Stochastic incipient motion criterion for spheres under various bed packing conditions. *J. Hydraul. Eng.-ASCE*, 128(4):369–380, 2002.

[125] E. Papista, D. Dimitrakis, and S. G. Yiantsios. Direct numerical simulation of incipient sediment motion and hydraulic conveying. *Ind. Eng. Chem. Res.*, 50:630–638, 2011.

[126] G. Parker and P. C. Klingeman. On why gravel bed streams are paved. *Wat. Resour. Res.*, 18(5):1409–1423, 1982.

[127] G. Parker and P.R. Wilcock. Sediment feed and recirculating flumes; fundamental difference. *J. Hyraul. Eng.-ASCE*, 119(11):1192–1204, 1993.

[128] C.S. Peskin. The immersed boundary method. *Acta numerica*, 11:1–39, 2002.

[129] R. Peyret and T. D Taylor. *Computational Methods for Fluid Flow*. Springer, 1983.

[130] S. B. Pope. *Turbulent Flows*. Cambridge University Press, 2000.

[131] M. R. Raupach and R. H. Shaw. Averaging procedures for flow within vegetation canopies. *Bound.-Lay. Meteorol.*, 22(1):79–90, 1982.

[132] W. C. Reynolds and A. K. M. F. Hussain. The mechanics of an organized wave in turbulent shear flow. part 3. theoretical models and comparisons with experiments. *J. Fluid Mech.*, 54:263–288, 1972.

[133] S.K. Robinson. Coherent motions in the turbulent boundary layer. *Annu. Rev. Fluid Mech.*, 23:601–639, 1991.

[134] W. Rodi. *Turbulence models and their application in hydraulics*. CRC Press, 1993.

[135] A. M. Roma, C. S. Peskin, and M. J. Berger. An adaptive version of the immersed boundary method. *J. Comput. Phys.*, 153:509–534, 1999.

[136] C. Santarelli and J. Fröhlich. On the pair correlation function in a bubble swarm. *Kerntechnik*, 78(1):50–51, 2013.

[137] G. Sardina, P. Schlatter, L. Brandt, F. Picano, and C. M. Casciola. Wall accumulation and spatial localization in particle-laden wall flows. *J. Fluid Mech.*, 699:50–78, 2012.

[138] A. Schoklitsch. *Über Schleppkraft und Geschiebebewegung*. Wilhelm Engelmann Verlag, 1914.

[139] G. Seminara. Fluvial sedimentary patterns. *Annu. Rev. Fluid Mech.*, 42:43–66, 2009.

[140] X. M. Shao, T. H. Wu, and Z. S. Yu. Fully resolved numerical simulation of particle-laden turbulent flow in a horizontal channel at a low Reynolds number. *J. Fluid Mech.*, 693:319–344, 2012.

[141] A. Shields. *Anwendung der Ähnlichkeitsmechanik und der Turbulenzforschung auf die Geschiebebewegung*. PhD thesis, Mitteilungen der Preußischen Versuchsanstalt für Wasserbau und Schiffbau, Berlin, (1936), (in German).

[142] A. B. Shvidchenko and G. Pender. Macroturbulent structure of open-channel flow over gravel beds. *Wat. Resour. Res.*, 37(3):709–719, 2001.

[143] A. Soldati and C. Marchioli. Physics and modelling of turbulent particle deposition and entrainment: Review of a systematic study. *Int. J. Multiphase Flow*, 35(9):827–839, 2009.

[144] A. Soldati and C. Marchioli. Sediment transport in steady turbulent boundary layers: Potentials, limitations, and perspectives for lagrangian tracking in DNS and LES. *Adv. Water Resour.*, 48:18–30, 2012.

[145] T. Stoesser. Large-eddy simulation in hydraulics: Quo vadis? *J. Hydraul. Res.*, 52(4):441–452, 2014.

[146] R. Storm. *Wahrscheinlichkeitsrechnung, mathematische Statistik und statistische Qualitätskontrolle*. Carl Hanser Verlag, Munich, Germany, 2007, in German.

[147] A.N. Sukhodolov and V.I. Nikora. Bursting and flow kinematics in natural streams. In R. Murillo, editor, *River Flow 2012*, pages 113–120. ISBN 978-0-415-62129-9, 2012.

[148] M. Uhlmann. An immersed boundary method with direct forcing for the simulation of particulate flows. *J. Comput. Phys.*, 209(2):448–476, 2005.

[149] M. Uhlmann. Interface-resolved direct numerical simulation of vertical particulate channel flow in the turbulent regime. *Phys. Fluids*, 20(5):053305, 2008.

[150] S. van Emelen, L. Schmocker, W.H. Hager, S. Soares-Frazão, and Y. Zech. Sediment transport models to simulate erosion of overtopped earth-dikes. In R. Murillo, editor, *River Flow 2012*, pages 499–506. Taylor & Francis, 2012.

[151] B. Vowinckel and J. Fröhlich. Simulation of bed load transport in turbulent open channel flow. *PAMM*, 12(1):505–506, 2012.

[152] B. Vowinckel, R. Jain, T. Kempe, and J. Fröhlich. Incipient motion of inertial particles in a turbulent open channel flow. In *ERCOFTAC Symposium on Engineering, Turbulence Modelling and Measurements. Marbella, Spain*, 2014.

[153] B. Vowinckel, T. Kempe, and J. Fröhlich. Impact of collision models on particle transport in open channel flow. In 7th *Int. Symp. Turbulence and Shear Flow Phen., Ottawa, Canada*, 2011.

[154] B. Vowinckel, T. Kempe, and J. Fröhlich. Particle-resolving simulations of bed load sediment transport. In 8th *Int. Conf. Multiphase Flow. Jeju, Korea*, 2013, paper No. 792.

[155] B. Vowinckel, T. Kempe, and J. Fröhlich. Fluid-particle interaction in turbulent open channel flow with fully-resolved mobile beds. *Adv. Water Res.*, 72:32–44, 2014.

[156] B. Vowinckel, T. Kempe, and J. Fröhlich. Highly-resolved numerical simulations of bed-load transport in a turbulent open-channel flow. In *NIC-Symposium, Jülich, Germany*,, 2014.

[157] B. Vowinckel, T. Kempe, J. Fröhlich, and V.I.Nikora. Numerical simulation of sediment transport in open channel flow. In R. Murillo, editor, *River Flow 2012*, pages 507–514. Taylor & Francis, 2012.

[158] B. Vowinckel, T. Kempe, J. Fröhlich, and V.I.Nikora. Direct numerical simulation of bed-load transport of finite-size spherical particles in a turbulent channel flow. In *DLES 9 (submitted)*, Dresden, Germany, (2013).

[159] B. Vowinckel, V. Nikora, T. Kempe, and J. Fröhlich. Momentum balance in flows over mobile granular beds: A double-averaging analysis of dns data. *J. Fluid Mech.*, submitted, 2014.

[160] M. S. Yalin and A. M. Ferreira da Silva. *Fluvial Processes*. IAHR/AIRH Monograph, 2001.

[161] Y. Yamamoto, T. Potthoff, M. Tanaka, T. Kajishima, and Y. Tsuji. Large-eddy simulation of turbulent gas-particle flow in a vertical channel: effect of considering inter-particle collisions. *J. Fluid Mech.*, 442:303–334, 2001.

[162] B. A. Yergey, M. L. Beninati, and J. S. Marshall. Sensitivity of incipient particle motion to fluid flow penetration depth within a packed bed. *Sedimentology*, 57:418–428, 2010.

[163] L. Zeng, S. Balachandar, and P Fischer. Wall-induced forces on a rigid sphere at finite Reynolds number. *J. Fluid Mech.*, 536:1–25, 2005.

A Buckingham Π-Theorem

As illustrated in Fig. 3.3 in Sec. 3.3.1, 10 dimensional parameters k_j with their physical dimensions of length in meter, mass in kilogram, and time in seconds describe the present problem. These are the free height above the sediment H, the kinematic viscosity of the fluid ν_f, the fluid density ρ_f, and the characteristic velocity of the flow, i.e. the bulk velocity U_b in terms of the channel flow, the wall shear stress τ_w or the characteristic friction velocity $u_\tau = \sqrt{\tau_w / \rho_f}$, $V_{tot} = L_x L_y L_z$ the total volume of the computational domain, and the the volume filled with fluid, i.e. $V_f = V_{tot} - V_p$ with V_p the total volume of the particles introduced in the channel. Furthermore, one obtains the submerged particle density $\rho_p - \rho_f$, the diameter of the particles D, the gravity g. In accordance with the Buckingham Π-theorem [23], these ten parameters listed above are summarized in Tab. A.1.

In order to apply the Buckingham Π-theorem to the present considerations, Table A.1 can be interpreted as a linear system of equations with the right hand side being equal to the zero vector [94]. Since there are only three basic physical dimensions, the matrix has the dimensions 3×10. To obtain the desired non-dimensional quantities, one has to find a 3×3 element of the matrix with its subdeterminant unequal to zero. This is indeed the case for the elements build by k_8, k_9, and k_{10}. Hence, the matrix can be written as

$$
\begin{aligned}
k_1 - k_2 + 2k_3 - 3k_4 + k_5 + 3k_6 + 3k_7 &= -k_8 + 3k_9 - k_{10} \\
k_2 + k_4 &= -k_9 \\
-k_1 - 2k_2 - k_3 &= 2k_{10} \quad ,
\end{aligned}
\tag{A.1}
$$

which simplifies to

$$
\begin{aligned}
-\frac{1}{2}k_1 - k_2 - \frac{3}{2}k_3 - k_5 - 3k_6 - 3k_7 &= k_8 \\
-k_2 - k_4 &= k_9 \\
-\frac{1}{2}k_1 - k_2 - \frac{1}{2}k_3 &= k_{10} \quad .
\end{aligned}
\tag{A.2}
$$

As a consequence, a non-trivial solution of (A.2) consists of seven non-dimensional quantities π_i that fully describe the physical problem, which is summarized in Tab. A.2.

The characteristic numbers that define the parameter space for the physical problem are computed as

$$
\pi_i = U_b^{k_{i,1}} \cdot \tau_w^{k_{i,2}} \cdot \nu_f^{k_{i,3}} \cdot \rho_f^{k_{i,4}} \cdot H^{k_{i,5}} \cdot V_{tot}^{k_{i,6}} \cdot V_f^{k_{i,7}} \cdot D^{k_{i,8}} \cdot (\rho_p - \rho_f)^{k_{i,9}} \cdot g^{k_{i,10}} \quad ,
\tag{A.3}
$$

and as a linear combination of π_i (Tab. A.3).

Dimension	U_b	τ_w	ν_f	ρ_f	H	V_{tot}	V_f	D	$\rho_p - \rho_f$	g
	k_1	k_2	k_3	k_4	k_5	k_6	k_7	k_8	k_9	k_{10}
Length $[m]$	1	-1	2	-3	1	3	3	1	-3	1
Mass $[kg]$	0	1	0	1	0	0	0	0	1	0
Time $[s]$	-1	-2	-1	0	0	0	0	0	0	-2

Table A.1: *Characteristic parameter of the problem of a particle-laden turbulent open-channel flow and their physical dimensions.*

Parameter	U_b	τ_w	ν_f	ρ_f	H	V_{tot}	V_f	D	$\rho_p - \rho_f$	g
	k_1	k_2	k_3	k_4	k_5	k_6	k_7	k_8	k_9	k_{10}
π_1	1	0	0	0	0	0	0	$-\frac{1}{2}$	0	$-\frac{1}{2}$
π_2	0	1	0	0	0	0	0	-1	-1	-1
π_3	0	0	1	0	0	0	0	$-\frac{3}{2}$	0	$-\frac{1}{2}$
π_4	0	0	0	1	0	0	0	0	-1	0
π_5	0	0	0	0	1	0	0	-1	0	0
π_6	0	0	0	0	0	1	0	-3	0	0
π_7	0	0	0	0	0	0	1	-3	0	0

Table A.2: *Matrix of the non-dimensional parameters π_i.*

pi_i	Dimensionless parameter	π_i^*	Characteristic number	Physical meaning
π_1	$\frac{U_b}{\sqrt{Dg}}$	$\pi_1 \cdot \pi_5^{-0.5}$	$Fr = \frac{U_b}{\sqrt{gH}}$	Froude number
π_2	$\frac{\tau_w}{(\rho_p - \rho_f)gD}$	π_2	$Sh = \frac{\tau_w}{(\rho_p - \rho_f)gD}$	Shields number
π_3	$\frac{\nu_f}{D\sqrt{Dg}}$	$\pi_1 \cdot \pi_3^{-1} \cdot \pi_5^{-1}$	$Re_b = \frac{U_b H}{\nu_f}$	Bulk Reynolds number
π_4	$\frac{\rho_f}{\rho_f - \rho_p}$	$\pi_2^{0.5} \cdot \pi_3^{-1} \cdot \pi_4^{-0.5}$	$D^+ = \frac{u_\tau D}{\nu_f}$	Particle Reynolds number
π_5	$\frac{H}{D}$	π_5	$\rho' = (\rho_p - \rho_f)/\rho_f$	Relative submerged density
π_6	$\frac{V_{tot}}{D^3}$	π_6	$\frac{V_{tot}}{D^3}$	Relative volume of the domain
π_7	$\frac{V_f}{D^3}$	$\pi_7 \cdot \pi_6^{-1}$	V_f/V_{tot}	Volume fraction

Table A.3: *Dimensional parameter π_i and the characteristic numbers.*

B Parallelisation and performance

The setups presented in sections 3.3, 4.2, and 5.2 consist of computational domain of several hundred million grid cells. It is obvious by the large number of grid cells that the present simulations can not be conducted on a single processor, but must be carried out on state-of-the-art supercomputers such as *Taurus* at the Center for Information Services and High Performance Computing (ZIH) [3], or JUQUEEN at the Jülich Supercomputing Centre (JSC) [4]. These petaflop computations usually include massively parallel applications with a large number of processors. A large amount of processors, however, requires a distributed memory strategy. In the code PRIME, this is realized by domain decomposition subdividing the total domain Ω into N_{proc} subdomains, with N_{proc} the numbers of processors employed. The communication between the subdomains is done by an explicit Message Passing Interface (MPI) using the library *Portable Extensible Toolkit for Scientific computations* (PETSc) [2]. Furthermore, the Helmholtz and Poisson equations described in Sec. 2.2.3 are solved using the library *High performance preconditioners* (Hypre [1]).

The particle data are also communicated by standard MPI routines employing a so-called *Master and slave* strategy [86]. The processor handling the sub-domain that contains the center of mass of the particle is called the master processor. To fully evaluate the equation of motion of the particles (2.3.2), the communication of the local forcing on the interface of the particle is needed. Hence, any processor handling a sub-domain that contains parts of the particle (but not the particle center) are considered as slave processors. This is illustrated in the simple example displayed in Fig. B.1. Here, the processor P_0 is the master processor for particle p, and P_1 is the master for q respectively. Moreover, P_1, P_4, and P_5 are the slaves for p. This strategy implies that for the case of a single particle, the maximal extent of particle reaching into a slave-domain is limited to the particle radius.

In addition, to properly account for particle collisions, exchange of particle velocities, both translational and angular, are needed to evaluate the equation for normal and tangential forces due to collisions, i.e. equations (3.5) and (3.6) respectively. Since the collision model, in particular the lubrication model (3.8), becomes active as soon as the gap between to particles is below $\zeta_{n,pq} < 2h$, the maximal extent of the area a particle influences becomes $R_p + R_q + 2h$ as indicated by the large dashed circel in Fig. B.1. Similar to the nomenclature discussed above, p_2 would be a slave particle for p_1 and vice versa for in the present example. This requirement, however, has implication on the domain decomposition, because in the code PRIME, the stencil for message passing is limited to neighboring subdomains, only. A communication between P_0 and P_2, e.g., is not possible and the slave particle must have its center of mass in one of the slave processors of the particle in question. Therefore, a proper domain decomposition yields a minimum local extent of a subdomain larger than $(R_p + R_q + 2\Delta_x) \times (R_p + R_q + 2\Delta_y) \times (R_p + R_q + 2\Delta_z)$.

The numerical efficiency of PRIME was investigated on JUQUEEN using up to 32768 cores. This was done by measuring the speedup $S(N_{proc}, N_{tot}) = T(N_{ref}, N_{tot})/T(N_{proc}, N_{tot})$ where

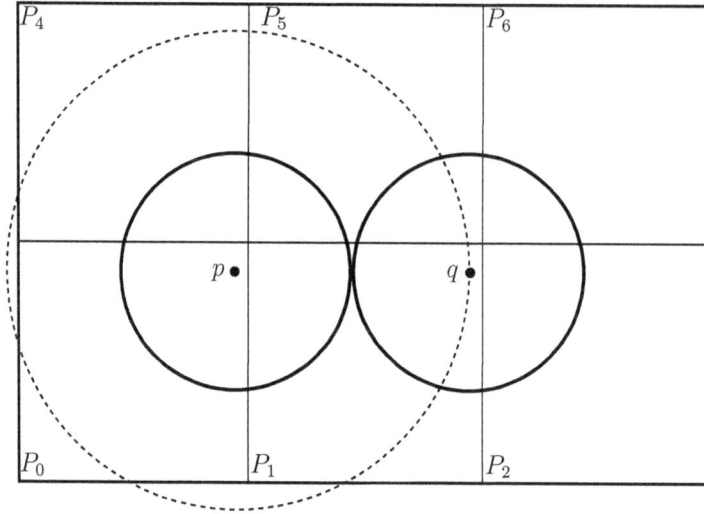

Figure B.1: *Sketch of the master and slave strategy of PRIME.*

N_{proc} is the number of processors, N_{tot} the total number of grid points, T the time to solve a given problem, and N_{ref} the number of processors used for the reference run.

As a first step, the speedup was measured for $N_{tot} = 268.5 \cdot 10^6$ grid points (Tab. B.1) in a rectangular channel $L_x/H \times L_y/H \times L_z/H = 16 \times 1 \times 8$. Here, the gravitational vector was oriented at a 45° angle to the wall and 12800 particles were randomly distributed across the computational domain. Note that this test case only addresses the initial time steps with the particles being evenly distributed across the computational domain. The reference run was carried out with $N_{ref} = 128$ processors and excellent speedup was observed despite the small amount of grid points per processor (Fig.B.2a).

The efficiency $E(N_{proc}, N_{tot}) = S(N_{proc}, N_{tot})/N_{proc}$ of the communication for message passing for this generic test case was tested by means of scale-up tests. In this case, the number of grid points per processor remains constant and the domain size is gradually increased (Table B.2). The scale-up with $N_{tot}/N_{proc} = 40^3$ shows good efficiency (Fig. B.2b and c). The deviations from the ideal scale-up stem from the random initial distribution causing different costs for communication. The scale-up $N_{tot}/N_{proc} = 51^3$ shows almost ideal efficiency.

The efficiency could only be tested for simulations with $N_{tot} \leq 1 \cdot 10^9$, because in PETSc the

Run	N_{proc}	N_{tot}/N_{proc}	absolute timing [s]	S_{ideal}	$S(N_{proc}, N_{tot})$	S/S_{ideal} [%]
Ref	128	$2.1 \cdot 10^6$	45.5	1	1.00	100
1	256	$1.05 \cdot 10^6$	23.3	2	1.94	98.0
2	512	$0.52 \cdot 10^6$	11.9	4	3.78	95.6
3	1024	$2.62 \cdot 10^5$	6.3	8	6.98	90.5
4	2048	$1.31 \cdot 10^5$	3.2	16	14.56	89.4
5	4096	$0.66 \cdot 10^5$	1.7	32	25.83	81.8
6	8192	$0.33 \cdot 10^5$	1.0	64	45.95	71.1

Table B.1: *Speedup with initial flow conditions for up to 8192 cores.*

	Run	N_{proc}	n_x	n_y	n_z	N_{tot}	N_p	absolute timing [s]
a)	Ref	128	256	128	256	$8.4 \cdot 10^6$	400	1.605
	1	512	512	128	512	$33.6 \cdot 10^6$	1600	1.698
	2	2048	1024	128	1024	$134 \cdot 10^6$	6400	1.672
	3	8192	2048	128	2048	$537 \cdot 10^6$	25600	1.800
b)	Ref	128	512	128	256	$16.8 \cdot 10^6$	800	3.137
	1	512	1024	128	512	$67.1 \cdot 10^6$	3200	3.142
	2	2048	2048	128	1024	$265 \cdot 10^6$	12800	3.183
	3	4096	2048	128	2048	$537 \cdot 10^6$	25600	3.194

Table B.2: *Scaleup configuration for a) $N_{tot}/N_{proc} = 40^3$, b) $N_{tot}/N_{proc} = 51^3$*

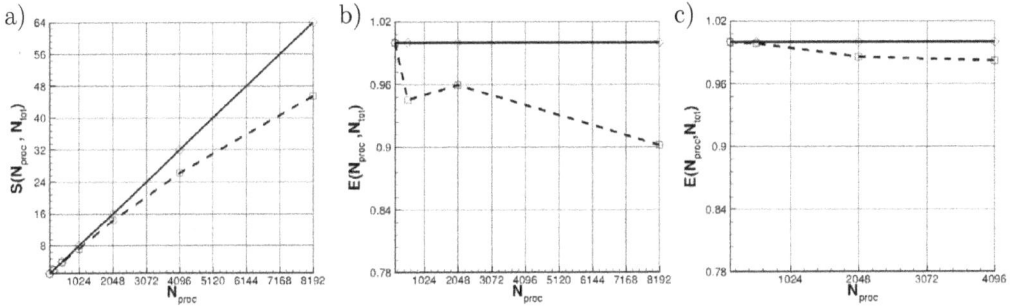

Figure B.2: *Performance of the code PRIME. a) Speedup with $268.5 \cdot 10^6$ grid cells and 12800 particles, b) scaleup for $N_{tot}/N_{proc} = 40^3$, and c) scaleup for $N_{tot}/N_{proc} = 51^3$.*

limit of 32-bit integers was reached. This configuration allows for $4.3 \cdot 10^9$ degrees of freedom (DOF). PRIME uses 4 DOFs per cell, i.e. the velocity components u, v, w and pressure p. For simulations with the number of grid cells exceeding this critical size, the data structure of the code was modified to one DOF, i.e. all velocity components as well as pressures are handled individually and a PETSc version with 64-bit indices was implemented. This now allows for up to $18 \cdot 10^{18}$ grid cells.

Subsequently, additional test for numerical efficiency of PRIME were carried on JUQUEEN using up to 32768 cores. Here the number of processors used for the reference run was $N_{ref} = 2048$. The speedup was measured for the developed flow field of case *Ref* presented in Sec. 5.2 covering 1.4 billion grid cells and 27000 particles. Due to their large density, particles preferentially accumulate on the bottom of the channel. Hence, an optimal domain decomposition would subdivide every process in a whole water column, i.e. only one processor in wall normal direction ($P_y = 1$) to guarantee an even distribution of particles among the processes. On the other hand, the master-slave strategy requires a stencil of subdomains with one subdomain being larger than the restriction discussed above, i.e. 22 grid cells in the present case. Therefore, a well balanced computation aims to maximize N_y/P_y and to minimize N_x/P_x and N_z/P_z respectively with N_x, N_y, and N_z the number of grid cells in the corresponding Cartesian direction and P_x, P_y, and P_z the number of processors subdividing the domain in the corresponding direction (Tab. B.3).

Excellent speedup was observed for the runs *Ref*, 1, and 2 using up to 8192 processors (Fig. B.3) confirming the results reported above for the initial situation of the simulations

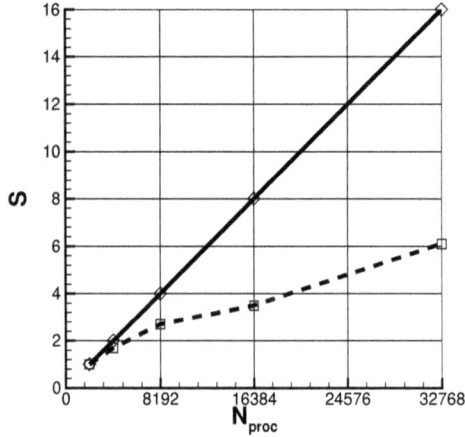

Figure B.3: *Speedup for the developed flow obtained from case* Ref *on JUQUEEN at JSC Jülich* (——: *ideal;* — — —: *PRIME*).

Run	N_{proc}	N_x/p_x	N_y/p_y	N_z/p_z	absolute timing [s]	S/S_{ideal}
Ref	2048	75	120	75	59.6	$1.0/\ 1 = 1.00$
1	4096	75	120	37.5	34.9	$1.7/\ 2 = 0.85$
2	8192	37.5	120	37.5	21.9	$2.7/\ 4 = 0.68$
3	16384	37.5	60	37.5	17.2	$3.5/\ 8 = 0.43$
4	32768	37.5	30	37.5	9.8	$6.1/16 = 0.38$

Table B.3: *Speedup for the developed flow obtained from case* Ref *on JUQUEEN at JSC Jülich.*

reported. For these runs, the criterion outlined above is optimized. Run 3 and 4 employing 16384 and 32768 processors use a larger amount of P_y due to the limits in domain decomposition in streamwise and spanwise direction. As a consequence, the speedup decreases and performance is lost due to the uneven computational load.

C Determination of the initialisation period

Every analysis of an unsteady numerical simulation must guarantee that the data used for the computation of statistical quantities are not influenced by the initial conditions. This task is solved here by defining a measure to determine the point in time, when the influence of the initial conditions has vanished and the statistically steady state is reached. In the case of a particle-laden flow, particle structures emerge with time scales much larger than the scales of the development of coherent fluid structures. Therefore, the measure developed here is based on the analysis of the particle velocity components. The identification of the initialization period is exemplified by the analysis of the linear velocity of the reference (case *Ref*) presented in Sec. 5.2.

The instantaneous first and second order statistics were calculated for the three components of the particle translational velocity vector (translational velocity in contrast to angular velocity). For this purpose, instantaneous data from time steps separated by one bulk unit were used. This is denoted by the operator $\langle ... \rangle$ indicating the average over all particles for an individual timestep. After the sedimentation to the rough wall, the mean wall-normal and spanwise velocity components, $\langle v_p \rangle$ and $\langle w_p \rangle$, respectively, reach a steady state within less than 5 bulk units (Fig. C.1). The streamwise component, $\langle u_p \rangle$, needs a larger amount of bulk units. This unsteady behavior becomes more obvious in the time signal of the standard deviations of all three components of the linear particle velocity (Fig. C.2) illustrating the need for an objective algorithm to determine the initialization period.

To investigate the significance of the unsteadiness of a particle quantity φ, here $u_{p,rms}$, $v_{p,rms}$, and $w_{p,rms}$, respectively, a linear regression analysis was performed to approximate the evolution by $\varphi_{reg} = a + bt$. This was done using the least square method (LSM) for the time interval $I_{reg} = [t_{start}, t_{start} + 48H/U_b]$. The length of the time interval, $48\,H/U_b$, is equivalent to two flow-through-times of the domain, which guarantees the inclusion of all relevant scales into the statistical fit. This interval also coincides with the time of the first maximum after the global minimum in Fig. C.2a. It can now be assumed that a stationary signal is developed, as soon as the trend expressed by the regression coefficient b is not significant anymore. The hypothesis, $H_0 : b = 0$, can be validated with a two-tailed t-test of the regression coefficient [146]. The standard two-tailed t-test evaluates the test statistics $|t_r^{(b)}|$ by comparing the variation of the signal s_r to the deviation from the calculated linear fit s_{t*} using the following quantities

$$t_r^{(b)} = \frac{b}{s_b} \tag{C.1a}$$

$$s_b = \frac{s_r}{\sqrt{(n-1)s_{t*}^2}} \tag{C.1b}$$

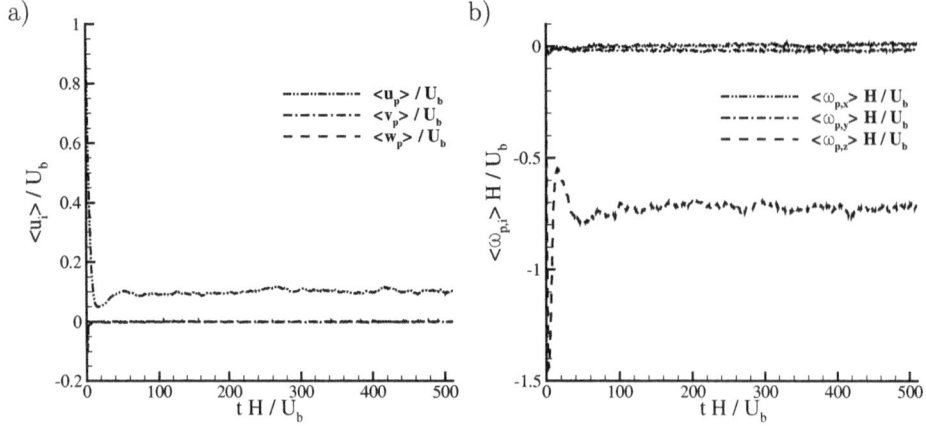

Figure C.1: *Volume-averaged components of the particle velocity over time of Ref. a) translational velocity, b) angular velocity.*

$$s_r = \frac{1}{n-2} \sum_{i=1}^{n} (\varphi_i - \varphi_{reg})^2 \tag{C.1c}$$

$$s_{t*} = \frac{1}{n-1} \sum_{i=1}^{n} (\varphi_i - \overline{\varphi})^2 \tag{C.1d}$$

with φ_i the discrete value of the physical quantity and $\overline{\varphi}$ the average of φ_i over $I_{reg}(t_{start})$. If $|t_r^{(b)}|$ is smaller than the critical value $t_{n-2,q}$ of the t-distribution for a given number of samples n and a given confidence interval q, the hypothesis cannot be rejected, i.e., the value of b is negligible and a stationary signal can be assumed. Due to the large number of samples (the criterion is $n > 50$), $t_{n-2,q}$ can be assumed to be constant. Hence, the critical value of a confidence interval of 95 % is $t_{n-2,0,95} = 1.67$. The results of the two-tailed t-test are displayed in Fig. C.3 for the three components of the components of the linear and angular velocity of the reference run. One or more of the calculated test statistics stay above the critical value of $t_{n-2,q}$ until $t_{start} = 25H/U_b$, i.e. a linear regression starting at $H/U_b = 25$ still yields a linear trend that is significant. A regression analysis starting at the critical time $t_{start}^{crit} > 25H/U_b$, however, shows a trend that is not significant anymore. To safely exclude the data influenced from the initial conditions, an initialization period of $t_{init} = t_{start}^{crit} + 48H/U_b$ was chosen. From this instant on, the observed quantities can be considered as statistically stationary. This type of analysis was performed for all simulations presented in the sections 4.2 and 5.2 to assure a high statistical quality of the data presented.

a)

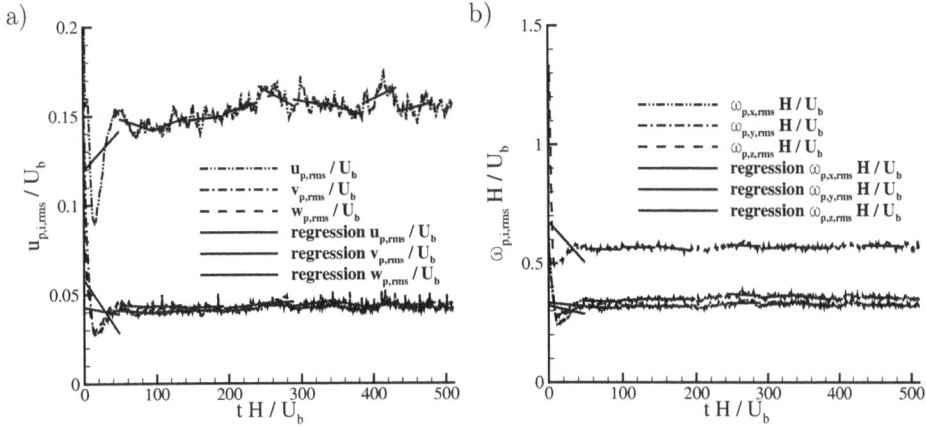

b)

Figure C.2:   *Standard deviation of the volume averaged particle velocity of Ref and its linear regression with ten intervals for illustration. a) translational velocity, b) angular velocity.*

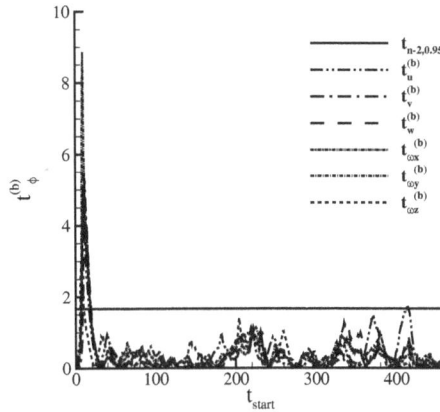

Figure C.3:   *Results of the two-tailed t-test of the hypothesis $H_0 : b = 0$ (case Ref) applied to the three velocity components of the particles u_p, v_p, w_p and the three components of the angular velocity $\omega_{p,x}, \omega_{p,y},$ and $\omega_{p,z}$.*

www.ingramcontent.com/pod-product-compliance
Lightning Source LLC
Chambersburg PA
CBHW081538220326

41598CB00036B/6476